RURAL NEW YORK

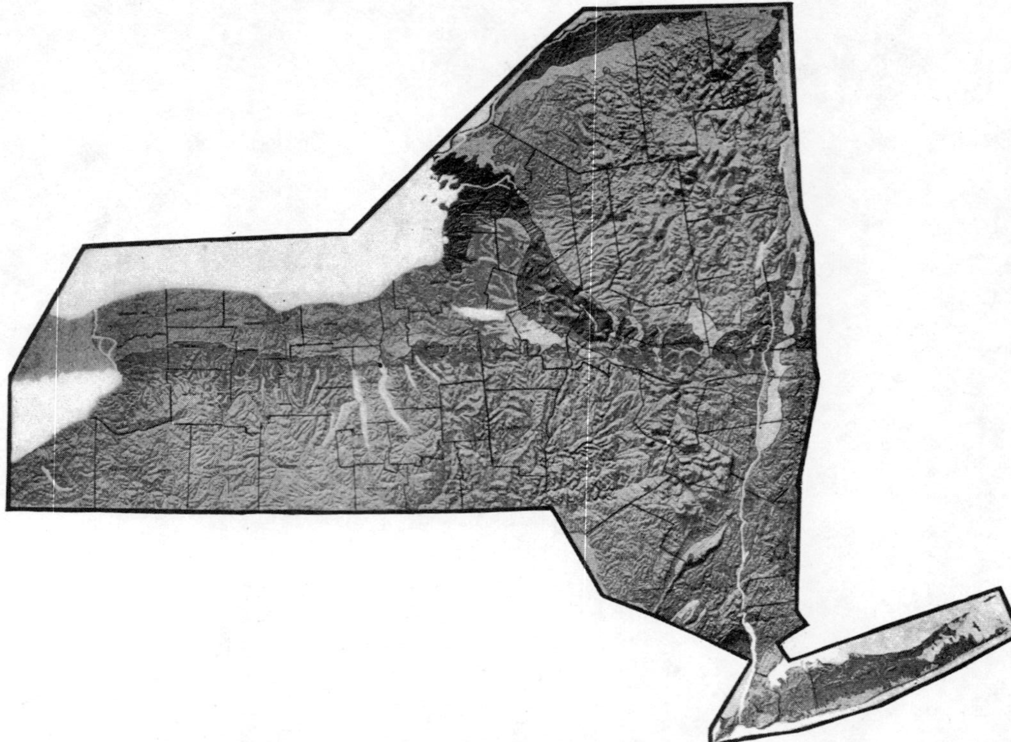

Plate I. Relief map of New York. The shading represents geological formations. (From model in New York State Museum.)

RURAL NEW YORK

BY

ELMER O. FIPPIN, B.Sc. (Agr.)
Extension Professor of Soil Technology in the
New York State College of Agriculture
at Cornell University

KENNIKAT PRESS
Port Washington, N. Y./London

RURAL NEW YORK

First published in 1921
Reissued in 1971 by Kennikat Press
Library of Congress Catalog Card No: 79-137943
ISBN 0-8046-1446-6

Manufactured by Taylor Publishing Company Dallas, Texas

KENNIKAT SERIES ON ECONOMIC THOUGHT,
HISTORY AND CHALLENGE

EDITOR'S PREFACE

Hereby is projected a new line of books to be known as the Rural State and Province Series. These books are to present the rural phase of the development of the commonwealths, with so much of the physical setting and history as will make plain the reasons for the present state of the agriculture and country life. The volumes are not guides to agricultural practice, nor books of advice; they are plainly descriptive. They are to be books of pleasant reading. At present some half dozen volumes are under contract. Should the public desire books of this character, it is the intention to include eventually all the States of the Union and all the Provinces of Canada.

The sources of information on the agriculture of New York, in its larger aspects, are not extensive in book form. The reader who desires to go farther than this little volume takes him will of course consult the reports and other publications of State departments and of the institutions, and the proceedings of societies. If he is interested in the history of the subject, he will profit by the four parts of the report of the "Society for the Promotion of Agriculture, Arts and Manufacturers," 1792–1799; the three volumes of the Board of Agriculture, 1821, 1823, 1826; and the reports of the State Agricultural Society 1841 to 1871, with subsequent less regular reports until 1899 and the semi-centennial volume of 1890, as well as the journal once published by the society. Historical matter is brought together in the "Cyclopedia of American Agriculture," particularly in Volume IV on education and institutions and very briefly in Volume I on John Johnston's work in tile-drainage; also in the editor's "York State Rural Problems," Volumes I and II, 1913, 1915 (J. B. Lyon Co., Albany). The files of the agricultural journals also afford invaluable material.

Unfortunately, the personal human documents are yet largely to be written. The reader will be interested in Roberts' delightful "Autobiography of a Farm Boy" (1916), published by the J. B. Lyon Co.

So these books go forth, the children of a long desire. I am grateful to the skilful authors who put the faith into the substance of the written word.

L. H. Bailey.

Ithaca, N. Y.
Sept. 1, 1920.

AUTHOR'S PREFACE

The aim of the author in preparing this volume is to present a bird's-eye view of the agricultural aspects of New York. He has endeavored not only to describe conditions as they are, but, when possible, to point out the underlying reasons for particular lines of development in agricultural production and rural affairs.

On the descriptive side, the book is the outgrowth of fifteen years of intensive study of the agricultural resources and development of the State. The approach has been made from the side of the soil. In the making of detailed soil surveys of different counties, in the continuous supervision of the surveys in close coöperation with the United States Department of Agriculture, in the giving of course of instruction in the New York State College of Agriculture on the soils and agricultural development of the State, and finally in almost continuous travel that has reached into all parts of the State, in by-paths and outlying regions as well as main paths of travel, these matters in all their interrelations have been the subject of continuous observation and study. Finally, through contact with a large part of the correspondence that has come over the greater part of this fifteen years touching soils and land improvement, the study and correlation of these agricultural resources and institutions have been ever-present problems.

Thus it is that this book is the fruit of those years of travel and study and teaching. Its preparation was in mind even before the specific opportunity for its preparation was offered by the Editor of this Series. The author welcomed the opportunity to join with the larger enterprise, and he desires to express to the Editor full appreciation for the many helpful suggestions that have been given as well as for the larger inspiration the author has derived from these years of official and personal contact. To many other friends and fellow-workers in and out of the State who have directly and indirectly contributed to the making of this book, he also desires to make full acknowledgment.

If this book contributes in some small measure to a better understanding and appreciation of the agricultural resources and institutions of New York in the larger sense, and the direction for their best development, it will have justified its preparation.

E. O. FIPPIN.

CONTENTS

CHAPTER I — PAGES

THE PHYSICAL AND CLIMATIC SETTING OF NEW YORK . . . 1– 55
General surface features . . . 4– 13
The structural geology . . . 14– 29
Drainage systems and lines of communication 29– 36
Physiography and industry . . . 37– 41
Climate of New York State . . . 41– 55

CHAPTER II

HISTORY OF AGRICULTURE IN NEW YORK . . . 56– 89
Agriculture of the Indians . . . 57– 60
Early settlement by the white men . . . 60– 64
Early routes and means of travel . . . 65– 69
Territorial development . . . 69– 75
Crop development . . . 75– 78
Development of under-drainage . . . 79– 81
The rise of agricultural institutions . . . 81– 89

CHAPTER III

SOILS OF NEW YORK . . . 90–117
Southern plateau . . . 94– 97
The Great Lakes plain and the Mohawk Valley 97–104
The St. Lawrence and Champlain valleys . . 104–108
The Hudson Valley region . . . 108–110
Adirondack Mountains . . . 110–111
Long Island . . . 111–114
Special soils . . . 114–117

CHAPTER IV

	PAGES
OTHER RESOURCES OF NEW YORK STATE	118–137
Fish and game	128–131
Mines and quarries	131–135
Water supply	135–137

CHAPTER V

AGRICULTURAL INDUSTRIES, PLANTS AND CROPS OF NEW YORK	138–199
Hay and forage	142–150
Pasture	150–152
Grain crops	152–161
Vegetables	161–172
Fruit	172–192
Nurseries and flowers	192–195
Special crops	195–199

CHAPTER VI

ANIMAL INDUSTRIES OF NEW YORK	200–244
Animal husbandry and soil fertility	209–212
Cattle and dairying	212–221
Horses	222–227
Mules	227–228
Sheep	228–235
Goats	235
Swine	236–238
Poultry	238–243
Bees and honey	243–244

CHAPTER VII

MARKETS AND MARKETING	245–270
Transportation and markets	248–257
Special marketing agencies	257–270

CHAPTER VIII

RURAL MANUFACTURES OF NEW YOTK	271–292

CHAPTER IX — PAGES

THE ADMINISTRATIVE AND REGULATORY ORGANIZATIONS OF NEW YORK 293–313
Political divisions and system of taxation . 294–297
Rural credit facilities 297–299
The Council of Farms and Markets . . . 299–309
The Conservation Commission 309–313
The Weather Service 313

CHAPTER X

THE EDUCATIONAL AND RESEARCH ORGANIZATION OF NEW YORK 314–354
Experiment stations 335–339
Extension work 339–346
Other educational organizations 346–354

CHAPTER XI

OUTLOOK 355–369
The approach to constructive policies . . . 359–369

INDEX 371

ILLUSTRATIONS

FIG.		PAGE
1.	Geological map of New York showing the distribution of the different rock formations based on their chronological relations. (After Tarr)	15
2.	Geological map of the State showing the distribution of the different types of rock formations. (After Von Engeln)	21
3.	Geological map of the State showing the general direction of movement of glacial ice as indicated by striations and the distribution of certain types of glacial formations. (After Von Engeln)	28
4.	Graphs showing by decades the total area in farms, the improved area in farms and the number of farms in New York	37
5.	Diagram showing the proportions of urban and rural population and the distribution of the latter between the village and the country in 1910	38
6.	Number and distribution of population in the State by place of residence	39
7.	Proportion of land area in farms by counties in 1910	41
8.	Value of farm real estate and buildings and the amount of expenditures specified	42
9.	Distribution of farms by size in 1910	43
10.	Showing the climatological districts in the State	44
11.	Showing the average number of days between killing frosts in spring and fall which corresponds to the average length of the growing season	48

FIG. PAGE

12. Average annual precipitation in different parts of the State 51
13. Average amount of precipitation in inches in the three summer months of June, July, and August in different parts of the State 52
14. Numbers and distribution of rural population of foreign stock by nationalities in 1910 64
15. Graphs showing by decades the total and the rural population of United States, the rural population, and the population engaged in agriculture in New York 73
16. Changes in rural population from 1900 to 1910 . 76
17. Showing the distribution of the main series and groups of soil 92
18. Total value and distribution of all crops and the amount and distribution of the specified crops in 1909 139
19. Amount and distribution of the specified crops in 1909 143
20. Graphs showing by decades the total area in crops and the area in hay and forage, in clover and in alfalfa 145
21. Graphs showing by five-year periods the area in corn for grain, total wheat, spring wheat, winter wheat, and oats 153
22. Graphs showing by five-year periods the area in buckwheat, barley, rye, apples, and tobacco . . . 154
23. Maps of the acreage and distribution of the specified vegetable crops in 1909 162
24. Graphs showing by five-year periods the area in potatoes, hops, dry beans, and cabbage 164
25. Maps of the amounts and distribution of specified fruits and crops in 1909 173
26. Amount and distribution of specified fruits in 1910 175

FIG. PAGE

27. Maps of the acreage and distribution of the specified fruits according to age in 1909 except for apples which are for 1910 177

28. Graphs showing by five-year periods the number of all cattle, dairy cows, work oxen, horses and mules and the production of butter and cheese . . . 206

29. Maps of the value, numbers, and distribution of the specified types of domestic animals in 1910 . . 208

30. Graphs showing by five-year periods the numbers of swine, sheep, and chickens and the production of wool and eggs 229

31. Maps showing number and distribution of sheep and swine in 1910 231

32. Number and distribution of poultry in 1910 . . 239

33. General map of New York State showing drainage systems, counties, and chief cities. Between pages 245 and 246

34. Diagram showing the organization of the Council of Farms and Markets into divisions and bureaus 300

35. Diagram showing the organization of the Conservation Commission into divisions and bureaus . . 310

36. Diagram showing the organization of the University of the State of New York and the numbers of units included in some of the main groups and divisions in 1915 316

37. Diagram showing the amounts and distribution of federal and state expenditures for agriculture in New York State 329

LIST OF PLATES

I Relief map of New York . . . . *Frontispiece*
(From model in New York State Museum)

FACING PAGE

II Model of Iroquois Indian house, stockade and family settlement 58
(From model in National Museum)

III Lettuce growing on muck soil 114

IV A grain field in the Genesee Valley 156

V Intensive fruit culture in Western New York . 190

VI Representative fruit plantations in New York . 224

VII Typical view in the hill country of Southern New York 284

VIII View in the Mohawk Valley region 346

RURAL NEW YORK

CHAPTER I

THE PHYSICAL AND CLIMATIC SETTING OF NEW YORK

THE State of New York forms the connecting link between New England and the remainder of the United States. It is the central one of the north-eastern group of states. It is the only state that touches the open water of both the Atlantic Ocean and the Great Lakes. In position, diversity and volume of resources, agricultural as well as industrial, in transportation facilities and population, New York is truly the Empire State. Its outlook for development in all these directions justifies the progressive spirit expressed in the motto on the great seal — Excelsior! — which may be translated roughly: "We aspire to higher and greater things."

New York stands at the northern extremity of that broad wrinkled ridge made up of mountains and plateaus popularly known as the Appalachian mountain system, that reaches up across the eastern United States from near the Gulf of Mexico. The lowest divide between the Atlantic Ocean on the east and the

Great Lakes and the upper Mississippi Valley of the Central West lies through the center of the State. At the eastern terminus of this lowest divide, between the east and the west, which is the chief thoroughfare for transcontinental travel, stands New York City, the metropolis of the American continent and the leading American port.

At the entrance to the magnificent harbor of the city, on Bedloe's Island, is the Statue of Liberty, which was a gift from the people of France to the people of America. It is at once the symbol of friendship, of equality, of opportunity, and a token of aspiration. It may well be called the symbol of American institutions and of the spokesmanship relation of New York State and City to the remainder of the nation.

Situated at the western extremity of this low divide on the eastern shore of Lake Erie and a short distance from the tremendous falls of Niagara, is the City of Buffalo, the second in size in the State. It guards the western gateway to this artery of travel and its development is based primarily on the industries and exchange that have grown out of the shifts in transport that naturally arise at this point, from the junction of lakes, canal and railroad conveyances.

The median latitude of the State is the forty-third parallel north, which is the same as Milwaukee in Wisconsin, Boise in Idaho, Vladivostok in Siberia, and Madrid in Spain. The median longitude is the seventy-sixth meridian west, about the same as that of Philadelphia and Cape Hatteras.

In outline the State is roughly an equilateral triangle, with the apex to the west. More accurately, the shape is that of a great shoe with the heel resting on the shore of the Atlantic at New York City, and trailed by Long Island. The toe rests against Lake Erie. Lake Ontario forms the instep and the St. Lawrence river reaches up the front. It has also been likened to a ship sailing westward, Long Island being the rudder.

The length of the State from east to west is approximately 335 miles, to which should be added the extension of Long Island beyond the main east line for seventy-five miles further, and lying south of Connecticut, making a total extreme length of 410 miles. The extreme breadth north and south is 315 miles.

Within these dimensions, New York State has a total area of 49,170 square miles, of which 1550 square miles are water surface. The net land surface is, therefore, 47,620 square miles or 30,476,800 acres. This is equal to the area of Ohio plus that of Massachusetts. It is four-fifths the area of all New England; about equal to that of North Carolina or Louisiana; a little less than the area of Iowa or Wisconsin, and one-third that of California. It is two-fifths the area of the British Isles, one fourth that of France, and three-tenths that of Japan, and considerably larger than the Island of Cuba with its depending islands.

The boundary is exceedingly irregular. In part it follows water courses but long stretches strike across-country on arbitrary lines having only an historical significance.

GENERAL SURFACE FEATURES

The surface features of New York State are very diverse, both in general and in detailed form. In their variety and form they make up some of the most beautiful scenery in the country. It does not reach the extreme of boldness found in the mountains of the West or of some other continental areas, but in picturesque outline, in variety of detail, and in cultural development, it charms the observer at many points. Mountain domes, broad valleys, undulating hills, lakes, waterfalls, fields, forest and stream, contribute to those elements that make up beauty and grandeur in the different parts of the State. This results from the long and complex geological history of the region which has developed a wide variety of rock strata and subjected these to severe denudation with unequal wearing and filling.

A three-armed system of main valleys centering at Albany and Troy divides the State into three main upland areas, a northern, a southern, and an eastern or southeastern area.

These main valleys are the Hudson-Champlain across the eastern side of the State and the Mohawk-Great Lakes valley through the middle of the State east and west. These, with Long Island, form seven rather distinct physiographic regions. Beginning with the lowlands they are: (1) The Hudson-Champlain; (2) the Mohawk Valley; (3) the Great Lakes Plains; (4) Long Island; (5) The Adirondack Mountains; (6) The Eastern Highlands; and (7) The Southern Plateau, which includes the Catskill Mountains at its eastern end.[1]

The highland and mountain areas with their rugged outlines are the more impressive in a scenic sense and they occupy the larger area. In the form of a series of great domes they reach a maximum elevation of 5397 feet in the peak of Mount Marcy in the Adirondacks and 4205 in the Catskill Mountains. The valley areas have a general elevation of 200 to 500 feet above sea level.

The Hudson Valley-Champlain region is a long and comparatively narrow trough, fifteen to twenty miles wide, rather sharply cut off from the Catskill and Adirondack areas by steep slopes but joined to the eastern highlands by a succession of hills that gradually increase in height so that there is no sharp line of division. It crosses the entire eastern front of the State. In the northern third along Lake Champlain, the valley is open to the eastward, so far as New York is concerned. While called a valley, it is not one in the sense of being a regular smooth erosion

[1] For a detailed view of the topography and cultural features of the State, the reader is referred to the topographic sheets that show the country on a scale of one inch equals one mile, with contours at twenty foot intervals, prepared jointly by the New York State Engineer's Office and the United States Geological Survey, and distributed by the Director of the latter at a small charge. These maps cover the greater part of the State and are by far the best geographic guides available.

valley. Rather it may be characterized as a low region but the surface is exceedingly irregular. It is a series of undulating to broken hills set apart by an irregular network of valleys more or less filled with glacial débris and in which most of the streams follow a tortuous and often sluggish course. These statements apply especially to the Hudson River section. The Hudson River itself occupies a very narrow, low-walled, mostly rock trench with almost no bottom land along its course. The Hudson gorge and valley extend out to sea in a southeast direction seventy-five miles beyond New York City. At the present time, the tide reaches up the river 150 miles to Troy. This situation is explained by the fact that in earlier geological times the region stood at a much higher elevation and the general system of valleys was then formed. Subsequently, during the glacial epoch, the region was depressed. Later, it was raised but not to its former position. Borings in connection with the development of the Ashokan water supply for New York City in the Catskills showed a trough eroded in the rock more than 1000 feet below the present surface of the Hudson River near Newburg, where the inverted svphon of the canal crosses the channel.

In the Palisades region below Newburg, the rocks rise for the most part precipitously from the water's edge. Farther north, that is half way to Albany, the rise is less abrupt and bold but still of a rugged hilly form. Beginning near the north line of Dutchess County and extending northward, a gently undulating plain fringes the river but at an elevation of sixty to one hundred feet above its surface. This is a clay plain with the exposed edge eroded into a system of steep slopes. It narrows and widens according to the position of the adjacent rock-cored hills. The plain is best developed above Troy and from there to a little above the great westerly bend of the Hudson River, in Washington and Saratoga counties, it is a broad plain, mostly capped by sand and gravel. From point to point the valley widens to receive some inflowing tributary, the largest of which on the west is the Mohawk River opposite Albany and Troy. On the east the most important tributary is the Hoosic River in northern Rensselaer County.

North of the "Great Bend" at Hudson Falls the general valley region is constricted to a very narrow pass by the east flank of the Adirondacks to unite with the Champlain Valley. For a number of miles there is only this narrow rock-walled lake, but further north near Ticonderoga the hills recede a little and place is given to a low and rather narrow lake-plain terrace of clay, with here and there sand and gravel shelves. The surface is undulating. The plain widens toward the north, though at places it is constricted. The greater width of lowland is on the Vermont side. In Clinton County the plains region blends with low hills that successively rise to the Adirondack Mountain system.

The Mohawk Valley which breaks out from the Hudson region at Troy is a V-shaped trough lying between the Adirondack and the Catskill mountains. Both masses approach the river by a succession of rounded hills of decreasing elevation. The interior trough is narrow and steep-sided but not to the same extent as the Hudson River. As far up as Little Falls, seventy miles from its mouth, there is only a narrow ribbon of alluvial soil a half to three-quarters of a mile in width, now on one side and now on the other, or divided. On the rolling hills on either side, the surface is smooth enough easily to permit successful farming for many miles back. The land rises less rapidly on the south than on the north side.

Above Little Falls, where a dike of intensive rock crosses the course of the river and gives rise to the falls that are the occasion for the city, the valley gradually widens out into a great amphitheater that looks out westward over Syracuse, Buffalo and the Great Lakes region. On the north the lowland swings gradually to the north around the western flank of the Adirondack Mountains, past the end of Lake Ontario where it forms a plains area ten to twenty miles in width with an undulating to flat surface. Thence it trends northeast along the St. Lawrence River and passes out into Canada. Opposite the source of the St. Lawrence in Lake Ontario, where the surface is in general low in elevation, there breaks forth a multitude of low ridges and knolls of solid rock, some of which constitute the bulk of the Thousand Island group. These are the outlyers of the Adirondacks. The Island region is a partially submerged valley. The soil is a combination of lake clays and sand and gravel terraces with some glacial till, interspersed between the projecting rock masses. The influence of the St. Lawrence Valley region is felt across the entire north line of the State where it joins with the Champlain Valley by a broad low ridge around the northern side of the mountains.

On the southern side of the Mohawk River the lowland area west of Little Falls continues with a slight southerly trend. The boundary past Utica is still quite abrupt but from there west it fades into a more gentle slope of decreasing elevation. West of Syracuse there is a uniform slope from the southern highlands to the shore of Lake Ontario. A little to the north of a line from Utica to Syracuse is a broad, flat plain in the midst of which lies Oneida Lake, a shallow sag in the land surface.

For a hundred miles west of Syracuse, which is two-thirds of the distance to Buffalo, the country forms a long, gentle, but undulating to hilly slope from the southern plateau region of the State northward to Lake Ontario. The six hundred foot contour is situated about thirty miles back from the lake and is followed a few miles further south by the one thousand foot contour. The surfaces of a large part of this rich plains region is made up of large, rounded and elongated hills, usually not over a hundred feet in height and with their long axes in the same general northerly and southerly direction. In fact, these tadpole shaped hills, which are most numerous in Wayne County and gradually decrease in numbers as one passes outward from that general region, are arranged concentrically and focus on a point on the northeast shore of Lake Ontario. While the slopes of these hills, known as "drumlins or hogsbacks," are usually steep, tillage is generally possible and they form a prosperous agricultural region. Between the hills is an intricate system of hollows often occupied by mucky swamps. To the south of the drumlin region, the land surface has a more general roll.

The southern margin of the plains region is notched at a half dozen places by rather narrow valleys occupied by lakes. These are the Finger lakes, the largest being Cayuga and Seneca. The country around the northern end of these lakes blends with the general plains region.

In its western part, the lake plains region breaks into two distinct levels separated by a steep slope or mountain. This escarpment is best developed in central Niagara County and crosses the Niagara River in the grand cascade of that river in its northward course between lakes Erie and Ontario. The maximum difference in elevation between the two plains is two hundred feet, at the Niagara River. The height of the main slope between these plains gradually decreases toward the east and disappears altogether a little west of Rochester. The lower or Ontario plain is distinctly flat, or slightly hollowed out near its southern margin. Its surface is relieved by a few low knolls and ridges. The upper or Erie plain is more complex. Near the shore of Lake Erie it is generally quite flat. This type of country swings south past Buffalo as a broad plain and then southwest along the lake. As it passes westward, the plain becomes more narrow and is progressively set off from the highland region by a sharp steep slope. At the west line of the State, the plain area is only a few miles in width. The remainder of the plains region, reaching well into the second tier of counties from Lake Ontario, is made up of rolling hills easy to farm.

Long Island should be associated with the up-state lowland region in the discussion of surface features. It is a long low ridge with smooth, and on the south side, low shores bordered by sand reefs and marshy tidal flats. The north shore is for the most part a steep bluff and is indented by a number of shallow bays. In general the elevation is from sixty to 150 feet above sea level but at several points in the interior there are rough hills, one of which reaches an extreme elevation of 380 feet. For the most part, the island is made up of a series of undulating sandy plains.

Considered as a whole, the three lowland regions embrace much the best agricultural and industrial parts of the State. Their total area, about 15,000 square miles, is approximately one-third that of the State. Exclusive of Long Island, which has an area of 1500 square miles, these lowlands include part or all of thirty-five counties. The high development in these regions is due to a fortunate combination of smooth topography, good soils, reasonably mild climate, and good transportation and market facilities.

The highland areas may be described more briefly. The boldest of these is the Adirondack Mountain district, which is made up of a succession of mountainous domes, culminating near the center of the area in Mount Marcy. Most of the area is timbered. The soils are thin and very stony. Small areas are cleared and tilled in the interior valley. Around the base of the mountain area is a narrow fringe of agricultural land that includes some prosperous farming districts, such as that in northern Oneida and central Herkimer counties.

The eastern highland takes in the western foothills of the Berkshire hills that begin on the eastern edge of the State and spread over into New England. It also includes the mountainous hills lying across the lower part of the State above New York City and the Palisades district of the Hudson River. The maximum elevation is not over 1500 feet but by contrast with the adjacent lowlands, it appears high and rugged and is distinctly mountainous and picturesque in outline. All the southern part of this area has been appropriated as a high-grade residence district.

The southern plateau is the most extensive of the three, as well as the most important agriculturally. As has been stated, it rises gradually to the south and east and in the latter direction culminates in the Catskill range. Near its middle part there is a sag of several hundred feet in maximum elevation, the western part being higher than the middle.

This vast plateau which is the northern margin of the Cumberland plateau, that reaches southward into Alabama, has been deeply eroded and now appears as a system of bold hills rising by gentle or steep slopes from the network of valleys that have been formed in its structure. While these broad valleys traverse the region in all directions, the prevailing one is north and south. They are continuous from one drainage system to the next and consequently form a system of deep passes, called Through valleys, that are the thoroughfares of modern travel. The Catskill portion being the highest, is most broken and has little agricultural importance. It occupies about a sixth of the area. The main plateau has a maximum elevation of 1800 to 2500 feet and is bounded on the north by about the 1000 foot contour.

The valleys, especially the north-south ones, have two main slopes: first, the interior, rather steep walled that bound the narrow valleys of alluvial and terrace land, usually of good agricultural value; second, the upper slope which forms a broad, moderately graded, U-shaped trough. Over most of these slopes, modern machinery may be used readily.

These distinctions in topography should be kept clearly in mind since they are essential to a correct appreciation of the agricultural and industrial development of the region later to be discussed.

A large part of the southern area is tillable land, especially that associated with good valleys, bearing prosperous farms. However, it is within this as well as the other highland areas, that the margin of profitable farming is most frequent and in which the shifts of economic pressure have had their most marked effects.

THE STRUCTURAL GEOLOGY (See Figs. 1–3)

The underlying geological structure of New York is very ancient. Its formations represent the dawn of the land areas of the American continent. Great mountain chains and plateaus have been formed and largely destroyed by erosion. New York lies on the west slope of the narrow ridge of land extending from New England to northern Georgia that formed the original backbone of the North American continent. For ages nearly all of the territory of New York was beneath the ancient ocean and received the succession of sediments and deposits washed from the upland to the east. The floor of this sea oscillated up and down unequally, somewhat like the billowy waves passing over a grain field, only vast periods of time were required for the movement. Sandy shores, mud flats, deep clear water and alkaline basins succeeded each other in their occupancy of the territory. The waste sediments from the upland came in from opposite directions and were, therefore, of different character. All these incidents are recorded in the great succession of rocks that now make up the structure of the State. With the exception of the Adirondack region, practically all the surface rocks are made up of marine sediments and accumulations, and their total depth is measured in miles rather than in feet. They begin in pre-Cambrian time, which was before the dawn of rock formations distinguishable by fossil remains of plants or animals. Vast successions of time are represented in rock strata that reach through the Lower and Upper Silurian and

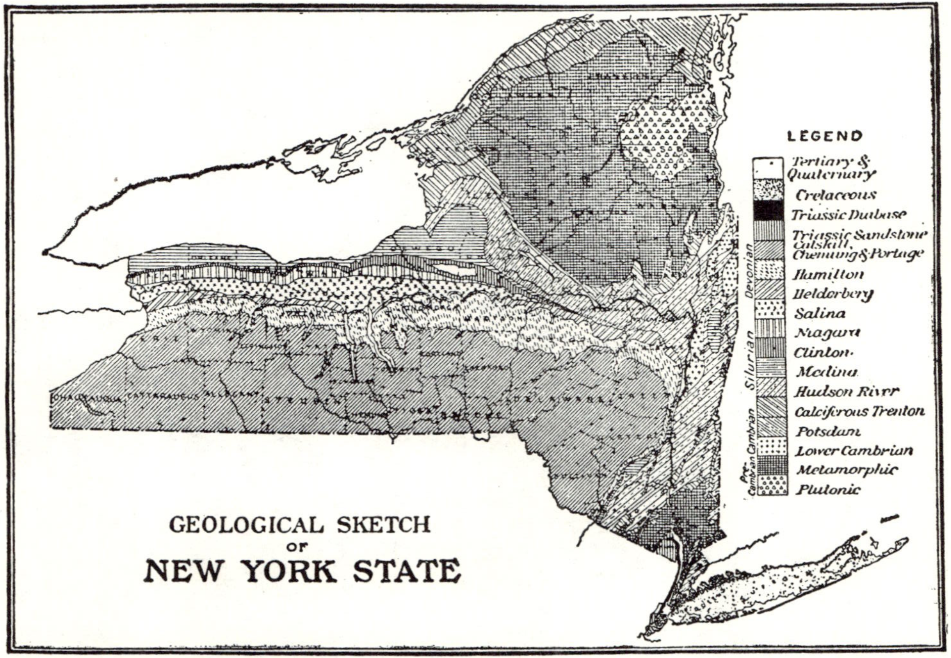

Fig. 1. Geological map of New York showing the distribution of the different rock formations based on their chronological relations.

the Devonian ages, up to the beginning of the Carboniferous or Coal age. Of this latter age the only representatives in the formations in the State are a few immense conglomerate bowlders in southwestern New York, of which the Panama rocks in Chautauqua County and at Rock City in southern Cattaraugus County, that were formed at the very beginning of that age, are the last remnants.

All through the long Coal age and to the present time, New York has been dry land with the exception of a small part above New York City which was submerged in the Triassic age following the Coal age.

For the most part, the land of the State has been well elevated and has been subjected to extreme erosion, although it is possible that later formations have been entirely swept away, as is suggested by the remnants of the lower coal measure rocks. In the vast ages during which the rock formations have been exposed to all the various destructive agencies, they have been eaten away and eroded by streams, the wind, and the waves, and carved into the main land forms, as they exist at the present day. The hard rocks resisted destruction more than the soft ones and form the eminences and rough slopes. The frail rocks were eaten into valleys and mild slopes. Thus the features of the State represent the scarred and weather-beaten products of ages of denudation acting on the succession of rocks of different degrees of resistence. Each gorge and waterfall and hill, if it could repeat its story, would give the succession of changes to which that region has submitted, and the present commerce and industry would be seen to conform to those geological changes. Of course it must be remembered that much has happened to the structure of the State since the rocks were formed and carved into the present general features. The changes due to the glacial epoch must be kept in mind as the immediate occasion of many features, but they have as their base and background the underlying rock structures. The soils will be seen to relate closely to all these formations and processes, so that in the study of the geological changes of a region one is laying the foundation for the understanding of the natural and human features of the states,— land forms, industries, commerce and sociology, and without this relationship one's conceptions of them must at best be very superficial.

The complex succession of formations and processes found in New York, together with the fact that it was one of the earliest fields of careful geological study in America, make it a landmark in progress. Since many of the American rock formations were first studied here, they have often become types and given local names that have been carried with those formations wherever they are found in America. Adding to this the monumental study of the "Natural History of New York," complete and published in five large volumes, and profusely illustrated, by special act of the legislature in 1841, one has some conception of the place New York occupies in the natural history annals of the country, and the basis it has for further development.

To consider the formations of the State in a little more detail, though briefly, one may begin with the oldest Precambrian areas. The Adirondack region is the larger of these, the southeastern highland above New York the smaller. Both have a core of plutonic rock centered by masses of granite. The higher peaks of the Adirondacks in Essex County belong to the granite group of rocks. Around this core of acid rock is a much larger area of the basic Gabbro rock, Norite, which skirts the base of the mountains and reaches across the St. Lawrence River near its source where it forms the Thousand Island group. This basic rock also extends well down to the Mohawk Valley.

In the second or southeastern area, the granite is associated with gneisses and schist rocks which probably represent the transformation of ancient sediments by pressure and heat. Together these rocks form the mountainous portion of that section. This region does not again become prominent in geological processes until in Triassic time when the bright red sandstones of lower Rockland County were formed coincident with similar formations through the Connecticut Valley in New England, and in various pockets throughout the Piedmont Plateau. At its close the great cracks in the structure permitted the intrusion from below of the molten material that now forms the Palisades rock along the lower course of the Hudson. The hard rock ridges that reach under the site of New York City are an important incident in its development, for they form the best kind of foundation for the ranks of tall buildings that distinguish New York City from Lonlon with soft marsh foundation. The channels that have been carved in the surface of this rock make the incomparable harbor that is also a large factor in the upbuilding of the city. These facts should not be forgotton in a sketch of the agricultural development and status of the State since the existence of such a city and port, with the market and transportation facilities it represents is a determinant factor in the agricultural development of the adjacent territory.

Touching again the Adirondack region, this broad dome is the center from which the succession of sedimentary rocks slope away in nearly ever direction. The up-turned or cut-off edges of those formations are arranged successively about the base of the plutonic core and form the floor of the country. The oldest rocks come first and disappear under the younger rocks as one passes out from the central mountain region. They dip to the south and west at a medium to slight angle. The older ones have been most eroded. The north-facing exposure of most of these rock outcrops has been an important fact in the succeeding development of the State because of its relation to the entrance of the Continental glacier and the subsequent history of the soil materials. There is on the north side a band of the Potsdam sandstone, a rather pure quartz rock, hard and lean in the elements of plant nutriment. Next, encircling the area is an irregular band of thick-bedded, high-calcium limestone, the Calciferous-Trenton. It touches the Mohawk River and passes under Lake Ontario. Its widest development is in central Jefferson County. On the south the next band of rock is a succession of gray to dark blue shales and sandstones known as the Hudson River group. They swing east over the southwest flank of the Adirondacks, along the Mohawk River, and thence southward through the Hudson Valley past the east slope of the Catskill Mountains. These rocks form the greater part of the floor of the Hudson Valley. Their eastern development has been much folded and contorted by mountain-making processes, so that throughout the Hudson Valley and progressively to the east line of the State the strata stand at a high angle and sometimes on edge. By this means bands of limestone were brought to the surface in a succession of pockets scattered through the valley. The metamorphosing forces of pressure and heat incident to the folding, changed the shale to slate, the sandstone to quartzite and the limestone to marble. All of these occur in deposits of commercial importance.

South of Lake Ontario the floor of the plain is formed by a quartzose sandstone and a gray or red shale interbedded with red sandstone. The sandstone is sufficiently hard to be used for paving brick and for building. This is the Medina, and reaches up to the foot of the "Niagara Mountain." In the Ontario Lake region of western New York many of the farm houses are built of smooth rounded water-worn bowlders or cobblestones of this rock that have been washed out of the glacial deposits and smoothed and rounded on the shore of the lake.

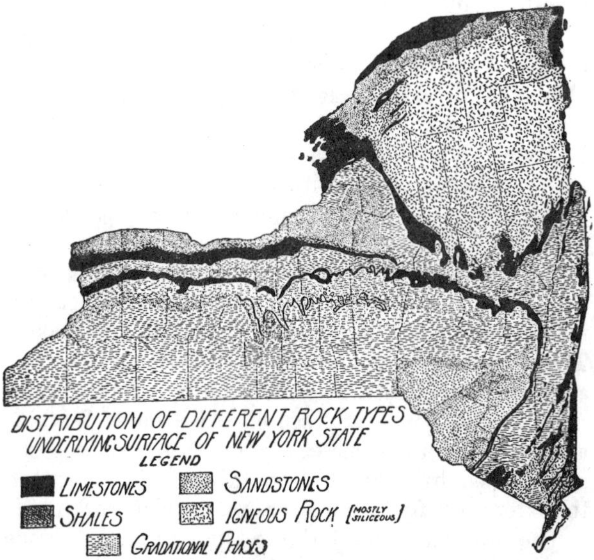

Fig. 2. Geological map of the State showing the distribution of the different types of rock formations.

Reaching eastward across the State from the Niagara River to a point south of Utica and forming the front of the Ontario escarpment, is a succession of limestone and drab shale formations. The lower of these, chiefly shale, is the Clinton formation on which Oneida Lake rests. Above it is the

Niagara formation capped by a hard magnesian limestone that forms the floor of the Niagara River and over which its cataract drops. This cap of hard limestone, underlain by the soft Lockport shale, constitutes the rock structure that gives rise to the magnificent Falls of the Niagara.

In the period succeeding the formation of the Niagara limestone, western New York was a shallow arm of the sea with a desert climate. In this sea there accumulated with the mud that now makes the Salina shale, great beds of salt and gypsum. It forms a band of country five to twenty miles in width as far east as Rome. At Syracuse, where salt springs originally occurred, salt is extensively manufactured, from which fact the name "Salt City" is taken. These springs were valuable to the Indians as well as to the early settlers and account for the beginning of the city of Syracuse. At Ithaca, Watkins and points in the Genesee Valley, these salt deposits are reached by wells and shafts extending through the later rock formations for commercial working.

The next and one of the less important limestone formations, the Helderberg limestone, lies above and south of the Salina. It is blue rock high in calcium content with thick bedding. Its outcrop is not wide but it reaches from near Buffalo, eastward around the east base of the Catskill Mountains. It passes by LeRoy, Geneva, Auburn, and has an even wider development in southern Oneida and Herkimer counties and the northern part of Schoharie County.

The position and extent of these various limestone

formations are of special importance because of their bearing on the later introduction of lime into the soils with its attendant agricultural significances. Their position with reference to the topography of the State and the important agricultural areas is also important in the matter of supplying lime to the soils that require it.

The remainder of the State upward and southward from the Helderberg limestone is a succession of shales and sandstones with occasional thin layers of limestone. Their color varies widely, but is predominantly gray or black. The first important division of these is the Hamilton shale which is somewhat calcareous. But the great mass of these formations is very lean in lime content and weather down to a light brown or yellow soil material. They belong to that part of the Devonian series usually known as the Catskill group. Some layers are hard sandstone, others soft shale. Capping the Catskills is a rather hard layer of conglomerate rock, expressed in the topography of the country as the protection of hills and the facing of steep slopes and waterfalls. Centering in Delaware County is an area of dark red shale and sandstone known as the Oneonta formation which has a marked effect on the soils of the region.

The topography of this southern region is the expression of the resistance of these soft and hard rocks to long erosion, coupled with the factor of elevation. The Catskills are highest, have the hardest capping and have been most eroded. Hence, it is a rough mountainous country of low agricultural possibilities.

The long period of erosion to which the State was subjected cut the surface into a very complete network of valleys. While the data are incomplete, the indications are that the drainage was mostly to the north and west to a great channel that occupied the axis of the Great Lakes and the St. Lawrence River. The topography of the southern highland was fairly rough. Through the middle and eastern part it was lower and in general more mild in outline than in other parts.

Into this long-eroded region there came the great continental glacial ice sheet. It advanced in a generally southerly direction from the Labrador highlands in eastern Canada. Slowly it advanced, pushing long tongues forward through the valleys. It was deflected by the land forms. Having filled the valleys, the ice advanced up the slopes until it covered all the mountain areas. Southward it advanced and piled deeper, until all the State was covered except a small angle south of the Allegheny River in Cattaraugus County. The rate of movement was generally in proportion to the depth and was most rapid through the valleys and slower over the hills. The old residual soil covering was swept away. Numerous eminences were planed off. Some valleys that crossed the course of the ice were partially or wholly filled, and those parallel to the ice movement were eroded still deeper, giving rise to "hanging valleys," which is the phenomena of side valleys high above the main valleys into which they flow.

The rocks in the course of the ice were ground up and carried in the direction of its movement, usually south. Material from various sources was mixed. Bowlders were mingled with finer material. Deep deposits were laid down in hollows, perhaps none on the eminences.

After a long and indefinite time, the ice began to retreat. Its margin fluctuated from season to season or from century to century. Where it stood for any length of time a ridge of earth was left and these terminal moraine-bands may be traced across the State in various places.

Great volumes of water were formed by the melting of the ice. This water accumulated in the hollows and valleys and flowed off to the southward as greatly flooded streams that formed broad bottoms of gravel and sand and fine loam.

As the ice margin retreated to the northward, it was, in all the western part of the State, moving down the general slope of the country. Consequently, the drainage water accumulated as ponds and lakes between the front of ice and the land divide in the valleys. As the ice retreated further north, these lakes found new outlets at lower levels. They connected through the valleys and finally came to have a large extent. Many successive levels of these are recognized in the more important valleys of the Finger Lakes region and the Genesee Valley, marked by the level of the benches of gravel and sand along their walls. At one stage this succession of lakes connected with the Lake Erie basin and drained into the Mississippi River through the Maumee River in northwestern Ohio. These were the Warren and Newberry stages of the lake. Gravel, sand, clay and other sediments were accumulated in these lakes, of which the gravel beaches on the Erie plain southwest of Buffalo on which the "Ridge road" is located are representatives. Later, as the ice retreated further north and the Mohawk Valley was cleared, a great lake was formed on the Ontario plain with its shore at the foot of the Niagara escarpment. This had its outlet eastward into the Hudson Valley and formed a great sand and gravel delta between Schenectady and Albany. This level of the lake built up a well defined sand and gravel beach that is readily traced from Lewiston on the Niagara River to Sodus on the shore of Lake Ontario. It is occupied throughout its course by an important public highway, another "Ridge road" lined by fruit-farms based in part on the materials that make up this beach. Other lake levels are recognized further north. The St. Lawrence Valley was flooded to a much higher level than at present. Likewise the Champlain Valley was occupied by a lake. A subsidence of the land in the eastern part of the State, due perhaps to the weight of the mass of ice, permitted the sea water to enter this valley at one stage. A glacial lake filled part of the Hudson Valley as far south as northern Dutchess County. In all these lakes more or less clay was deposited where the water was deep and quiet. These differ from other clays in being more or less strewn with bowlders, probably carried in on floating masses of ice. Probably the lakes were not deep. There were islands of glacial till formed from the mass of rock débris left directly by the ice as it melted.

Long Island, which has a core of Cretaceous sediments, was also covered by the ice and has a rough moraine along the northern shore and another diagonally through the central part from Brooklyn to Montauk Point. On the south side the surface was depressed below the sea during the retreat of the glacial ice and during that period broad plains of sand, sandy loam and clay were formed from the glacial outwash of New England.

The net result of the glacial incursion was to deface the preglacial surface features, rub off the sharp eminences, unevenly fill up many of the valleys and hollows with till and sediments and to carve out others perhaps a little more deeply. On the plains area south of Lake Ontario the till or rock flour from the grinding action of the ice was left in rounded bunches that in some places conform to the general rock surface and in others make tadpole shaped hills, called drumlins. These latter are widely distributed through the middle of the State but are especially abundant west of Syracuse and north of Auburn and Geneva. Wherever the general surface was moderately flat, irregular undulations were produced in which lakes and ponds were formed, many of which are now marked by swamp areas that are made up of deposits of peat, muck and marl. Through these, streams meander sluggishly in a very winding channel.

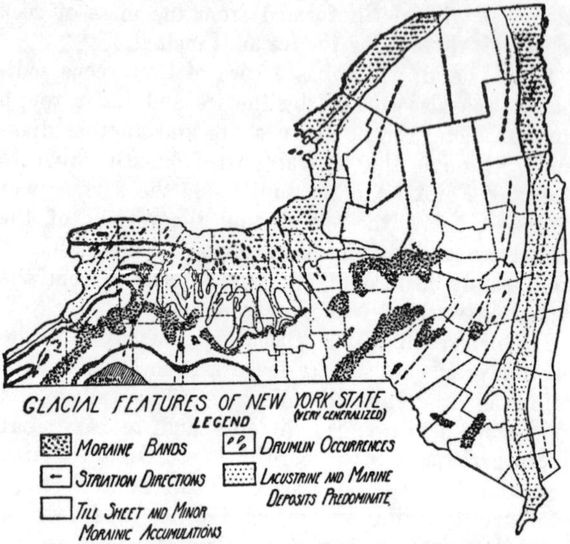

Fig. 3. Geological map of the State showing the general direction of movement of glacial ice as indicated by striations and the distribution of certain types of glacial formations.

The course of all streams was changed more or less. Some were sent over the edge of their old valleys and have cut deep gorges in the rock walls and frequently important waterfalls have been produced that have been the occasion for the location of more than one settlement that has grown into a second or third class city. Rochester and Little Falls are the most notable examples of this effect.

Agriculturally, the glacial epoch had a profound influence by its effect on the soils and on the lines of travel. Mostly, it improved by producing better soil and by smoothing the grades. A few grades it steepened and gave a certain minor unevenness. Such are the rises out of the deep north-south valleys and the over-hill roads through the drumlin regions. Over the western part of the State, a general mantle of till and rock débris was left and rock exposures are infrequent. In the northern and eastern parts where the rock relief is more bold and where the formations are more resistent, the earth covering is not so general and rock exposures are common and cut the country into irregular areas that are difficult of agricultural development, especially in this latter day when operations may be carried on with large machines.

DRAINAGE SYSTEMS AND LINES OF COMMUNICATION

New York State is well watered and well served by waterways. It has ocean frontage, an extensive shore line on two of the Great Lakes, many large and small interior lakes and navigable and important rivers, and other channels connecting the best parts of the State with the ports of the world.

The gateway to the ocean is at New York City where, although the effective shore-line is short, the harbor facilities are remarkably good. The long shore-line of Long Island is generally low and flat and does not afford good ports for large vessels. Of course ships may pass on the Sound as well as through the "Narrows" between Long and Staten islands. The partially submerged rock gorge of the Hudson River, with the inundated side channels that form the East and Harlem rivers, make a sheltered port on deep waterways having an unusual length for a single city. At the same time, the low gradient of the upper course of the river with the lack of large tributaries that are active in erosion permit the maintainence of the harbor in first class condition for the largest boats and this is accomplished with the minimum of labor. The largest ocean vessels find ready wharfage in the heart of the Greater New York district and may readily pass many miles up the Hudson River. At the same time, the land-locked harbor guarded across the Narrows by Long and Staten islands affords safe protection from the ocean storms.

On the Great Lakes the State has a total frontage of approximately 275 miles, one-fourth on Lake Erie and the remainder on Lake Ontario. The elevations of the surface of these lakes above sea level are 573 feet and 247 feet respectively. The harbors on either lake are not especially good. Buffalo is the port of entry on the former, Oswego and Rochester on the latter. By means of the Welland Canal across the neck of the province of Ontario, Canada, between the two lakes, large vessels may pass between all the Great Lakes and out to sea by the northern route through the St. Lawrence River. From the St. Lawrence River, medium-sized boats may also have access to Lake Champlain and by canal to the Hudson River.

There are numerous interior lakes, mostly small ones, the most important group being the Finger Lakes in the deep north-south valleys in the western part of the State. These are long and narrow and usually of great depth, the bottom of Cayuga and Seneca lakes reaching below sea level. Oneida Lake a little to the northeast of the Finger Lakes is the only other important interior body of water. Most of these lakes have been useful as means of travel and transport, having been connected up with streams and canal systems to form lines of communication. Their place in this service is not now as important as formerly, before railroads had been extensively developed. In the Adirondack Mountain region, there are hundreds of irregular lakes hid in the hollows and forests and utilized only for fishing and summer resorts.

The drainage of New York State falls into three main divisions: the Atlantic through the Hudson, Delaware and Susquehanna rivers; the St. Lawrence through the Great Lakes and the Champlain; and the Mississippi which embraces only a small territory in the southwest part of the State, represented by the Allegheny system. Of these streams the Hudson is the only one that is navigable to an important extent and on this large boats may pass as far as Albany.

The most important tributary to the Hudson is the Mohawk River that occupies the broad valley between the Adirondacks and the Catskill Mountains and whose course forms part of a very important thoroughfare of travel. Both the Hudson and Mohawk Rivers head in the southern foothills of the Adirondacks, and receive the drainage from the north and east sides of the Catskill Mountains and from the slopes east of the Hudson, including the Hoosic River as the largest tributary.

The St. Lawrence River receives the drainage of the northern third of the State, including that which flows into the Great Lakes.

The drainage of the southern highland is mostly reversed from that of the general slope of the country and instead of passing northward is carried to the south into the Delaware, Susquehanna and Allegheny systems. The exception is the Genesee River in the middle western part which cuts entirely across the State from south to north and empties into Lake Ontario above Rochester. A little drainage also goes out in that direction through the Finger Lakes.

The headwaters of the three south-flowing rivers mentioned are located on the very brink of the upland overlooking the Great Lakes plains. The water within three or four miles of Lake Erie flows south into the Mississippi system. These main lines of drainage guided the course of early emigration and the overland paths between drainage systems are marked by villages and cities, such as Rome and Buffalo.

The course of the streams, small and large, is irregular in direction and grade. The glacial influence superimposed on the general rock topography has made them a succession of sluggish winding channels and plunging rapids or cataracts. Waterfalls are common and are extensively employed. There is a large development of these falls at Rochester, Little Falls, Glens Falls and Watertown, and, of course, at Niagara Falls. Numerous small falls are used and in addition there is a large horse power in waterfalls not yet utilized. Among these may be mentioned that on the larger streams flowing out from the Adirondack Mountains and that of the Upper Falls of the Genesee at Portage, where there is the opportunity to construct a large storage dam to equalize the flow of water, and at the same time relieve the lower channel of the water that frequently causes disastrous floods. It has been proposed to develop electric power from these falls and distribute it under State superisvion to farms as well as to city enterprises.

Further, the State abounds in springs and one of the features of its hospitality is the wayside watering-trough on almost every highway, perpetually supplied from these sources. Incidently, it may be mentioned that a certain amount of tax is remitted to each land owner who maintains such a convenience on a highway.

New York State has an extensive canal system now largely fallen into disuse, with the exception of the one great thoroughfare, the Barge Canal, successor to the Erie Canal, and its tributaries which the State is just now completing. The total mileage that has been constructed is about 800, of which about 530 miles is now in service. The greater part of this is represented by the Barge Canal which connects the Hudson River at Troy with Lake Erie at Buffalo through the Mohawk Valley and across the Great Lakes plain. It utilizes the channel of rivers and lakes where possible. Extensive locks are required. Branches also connect it with Lake Ontario at Oswego, and with Cayuga and Seneca Lakes at the upper ends of which are terminal facilities. The Erie Canal was first finished in 1826, enlarged in 1840 to a depth of seven feet with corresponding width and has just now been entirely reconstructed and further enlarged to a twelve-foot channel to carry large barges, under the name of the Barge Canal.

The other important canal is the Champlain which connects the navigable portion of the Hudson River at Troy with the foot of Lake Champlain through the pass by Fort Edward, Fort Anne and Whitehall. This also has been enlarged to Barge Canal dimensions.

Other important canals that should be mentioned because of the part they played in the early development of the country are the Delaware and Hudson Canal, connecting the Hudson River at Rondout with Honesdale, Pennsylvania, which was opened in

1828; the Chenango Canal, connecting the Mohawk River at Utica with the Susquehanna River at Binghamton, completed in 1837; and several shorter canals acting as feeders between the Erie Canal system and the river systems on the south boundary of the State.

In the days before the advent of railroads, when the country was still new, the completion of these cheap means of travel and freighting were a strong impetus to agricultural production and development. The course of nearly all the canals is now paralleled by railroads, some of which follow the old tow-path of the canal. By their mobility, speed and the possibility of reaching outlying sections not practicable to be reached by water routes, the railroads have largely supplanted the canal systems and give better and more complete service for any but the most heavy and slow-moving freight, such as lumber, grain, ore and fuel.

A glance at any modern map of the State will show the extent and the general position of the railroads. In 1915 there were 8550 miles of steam road and 5000 miles of electric road. The great part of this mileage is coördinated into a half dozen great systems of regional or transcontinental spread. Thereby, through travel is greatly facilitated. The real course and position of these roads is not so apparent from the ordinary map. The topography of the State has exerted a large effect on them and has given to one place and taken from another where their natural advantages were otherwise equal.

The big factor in efficient and cheap lines of travel is the grade and the elevation to be overcome. Every pound raised and every foot it is raised requires a corresponding amount of power. This means the consumption of fuel as well as expense in equipment. Consequently, lines of travel seek the courses of least resistance. They keep to the valleys or climb the hills part way to dodge over a low divide or through it in a tunnel to the next valley. The direction and grade of the valleys determine their availability for transport. Reference to the discussion of the topography of the State will point out the practicable lines of travel. The same topographic features that hamper travel have interfered with agricultural and industrial development, and hence have put a further handicap on their expansion. All the more important valleys, as well as lines of connection across the plains, are occupied by railroads. The electric road with its still greater elasticity is reaching places not practicable for steam roads and the motor bus and truck are still further extending the service to remote regions. The prevailing north and south course of the valleys in the highlands makes cross country travel in the other direction circuitous and difficult. Further, the first slope from the valleys is likely to be quite steep which greatly limits the load and the speed of travel. The development of the State and the extension of particular enterprises, especially agriculture, must be with due regard for these physical limitations.

PHYSIOGRAPHY AND INDUSTRY (See Figs. 4–9)

The distribution of the population of the State is an unconscious conformity to these natural forces.

The great cities have grown up at the gateways where are the big exchanges, transfer, trade and industry, at New York and Buffalo. In the valleys or

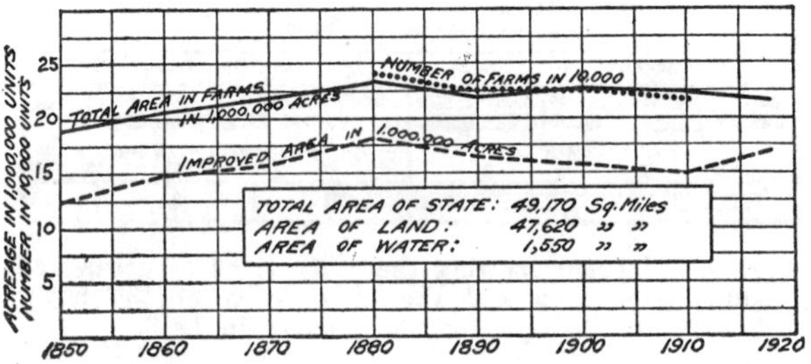

Fig. 4. Graphs showing by decades the total area in farms, the improved area in farms and the number of farms in New York.

on the plains at the more productive centers are the intermediate cities, as Albany, Utica, Watertown, Syracuse and Rochester. A little further out in the side valleys that are not main lines of travel is the second class city of Binghamton, and a number of third class cities such as Elmira, Ogdensburg, Plattsburg, Auburn, Ithaca, Cortland and Geneva; and finally there are the intermediate small towns and villages that set in the side valleys and a little back on the slopes gathered about first by a narrow ring of active farming region catering chiefly to the local needs; then the wider ring of less active farms that reach out to the remote hills and rough valleys on which latter is found the margin of profitable farming and industry. This most remote region is the fringe that is frequently frayed by the whippings of economic and social pressure.

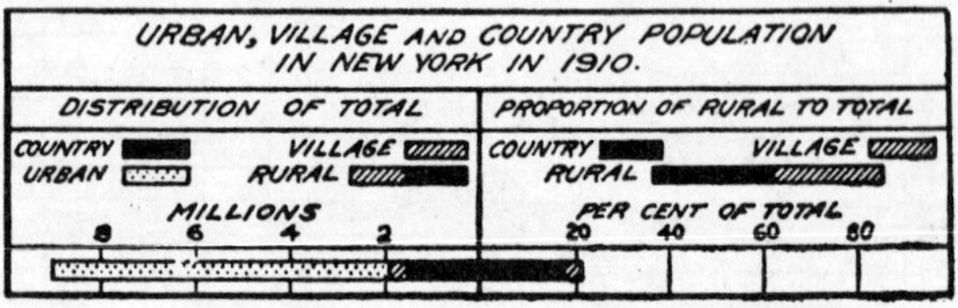

Fig. 5. Diagram showing the proportions of urban and rural population and the distribution of the latter between the village and the country in 1910.

Roughly stated, about 71 per cent of the land in the State is in farms. The remainder is mostly in mountains and forests. An appreciable area is occupied by cities, towns, villages, roads and similar cultural features.

Of the area in farms, 22 per cent is in woods. Over half of this is also used for pasture. Including the woodland, 36 per cent of the farm land is in pasture. Of the cleared land in farms, 57 per cent is in hay and pasture. About 59 per cent of the farm area is tillable.

An analysis of the rural population which constitutes only about 20 per cent of the total population of the State shows the following distribution:

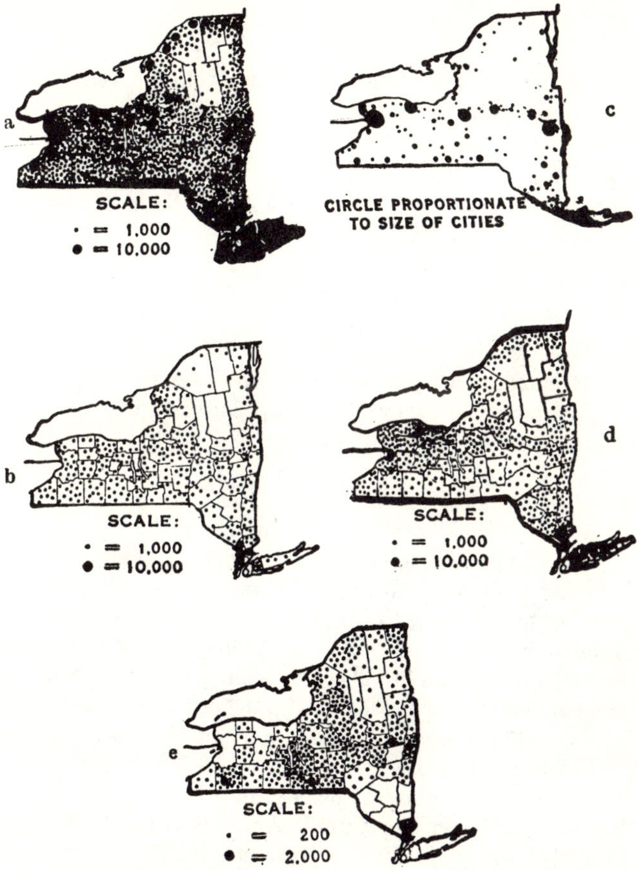

Fig. 6. Number and distribution of population in the State by place of residence: a. country; b. village; c. city; d. country population of foreign birth; e. decrease in population 1900 to 1910.

Table I. Distribution of Rural Population

1910 Total rural population	1,928,120
1910 Total country population	1,575,826
1917 Total farm population	830,517
1917 Total men and boys over 14 years	339,019
1917 Total women and girls over 14 years	271,333
1916 Number of hired men	104,332
1917 Number of hired men	88,310
1918 Number of hired men	47,433
1917 Percentage of farms having hired men	21.1
1918 Number of farms	200,903

While the figures for 1910 and 1917 are not entirely comparable, yet they are closely related.

This tabulation shows that 43 per cent of the rural population actually live on farms, and that about 40 per cent of that farm population is men and boys fourteen years of age or older. There is an average of 1.7 men and boys to a farm and as the average size of farm is 95 acres, the average amount of land handled a man is 56 acres. However, the average number of crop acres a farm is only 41 acres and the number of crop acres handled a man is 24 acres. By adding to the number of men the number of women and girls who work in the field, the crop acreage a person would be reduced to about 20 acres.

The rapid decrease in the number of hired men on farms in recent years indicates the difficulty of the labor problem on the farm. An indication of the unsettled social conditions is found in the proportions of males to females on farms, which in 1917 were 1.2 and 1.

A relief map of the State representing the lowest elevations by the darkest color and the highest elevations by successively lighter shades would to a very large extent represent the distribution of population, modified a little by the artificial means of communication.

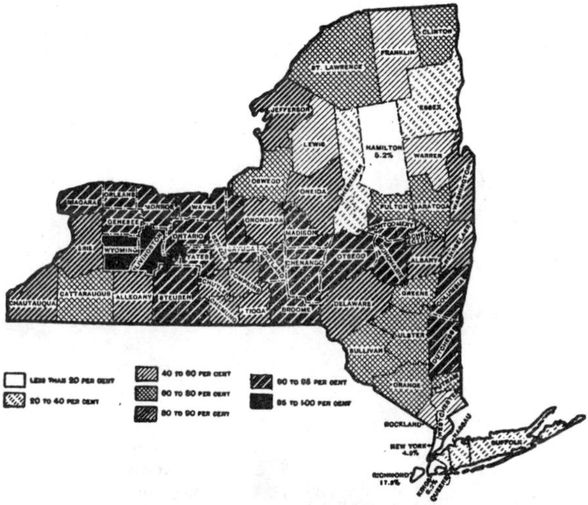

Fig. 7. Proportion of land area in farms by counties in 1910. Average percentage for the entire State 72.2.

CLIMATE OF NEW YORK STATE (See Fig. 10)

The climate of New York exhibits wide variations in all its elements. These result not only from the size of the State but also from its elevation, its variation in surface features, its position with reference to large bodies of water, the Atlantic Ocean and the Great Lakes, and finally from its position on the

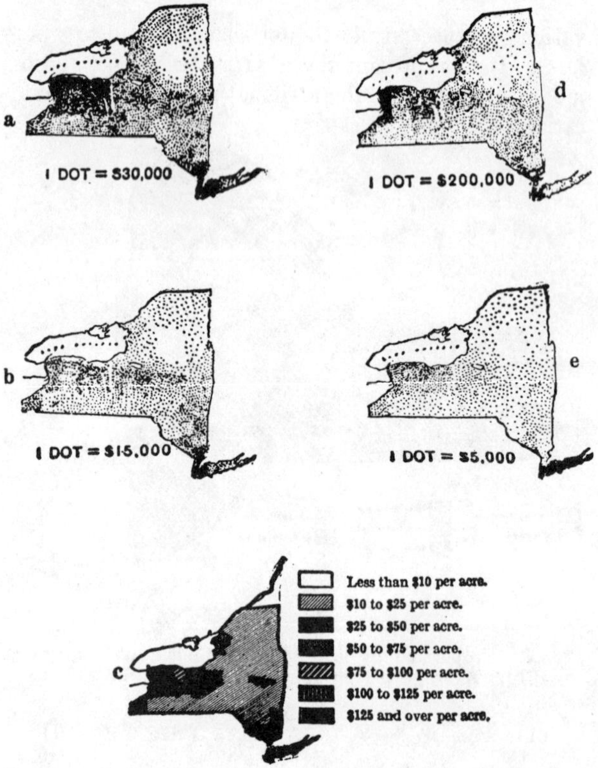

Fig. 8. Value of farm real estate and buildings and the amount of expenditures specified: a. value of buildings; b. expenditures for labor; c. average value of farm land an acre; d. value of farm implements; e. expenditures for fertilizers.

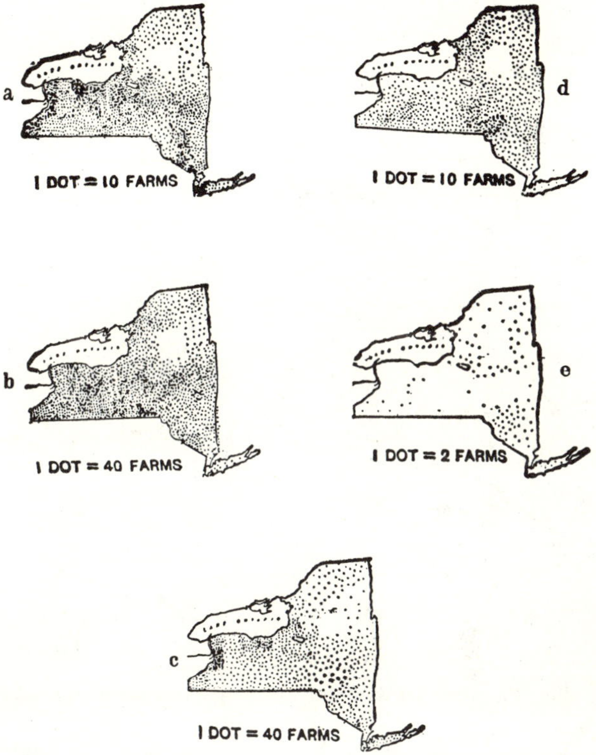

Fig. 9. Distribution of farms by size in 1910: a. 3 to 9 acres; b. 20 to 49 acres; c. 100 to 174 acres; d. 260 to 499 acres; e. 1000 or more acres.

continent with reference to the general course of the cyclonic areas that pass across the country in more or less regular succession. As a result of these conditions, it exhibits the three main types of climate: (a) the marine on Long Island and the southeast

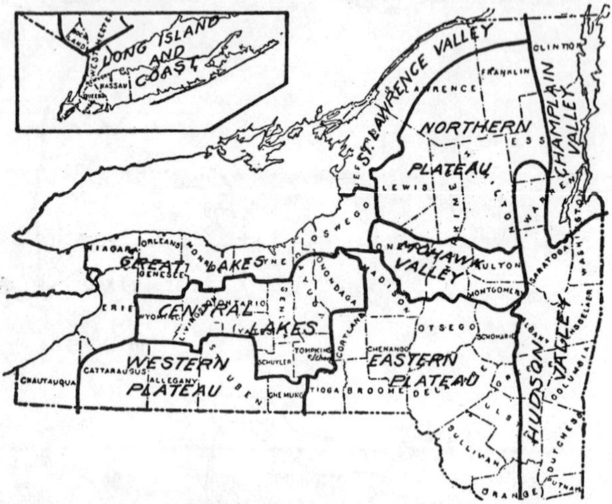

Fig. 10. Showing the climatological districts in the State.

point of the State; (b) the continental on the southern New York plateau; and (c) mountainous climate, especially well developed in the Adirondacks.

On the basis of the direction of wind movement and temperature and distribution of the precipitation, the State may be subdivided into ten divisions, the limits of which are shown in Fig. 10. The first organization for local climatic observations in America was that inaugurated by the New York Board of Regents in 1826 in coöperation with numerous academies under their supervision, which arrangement was continued until 1863. Shortly after that date, other agencies took up the work and continued it with some degree of completeness until 1900, when the National Weather Service was inaugurated with which the State service was affiliated.

The passage of the storm centers through the Great Lakes region is accompanied by shifts in the direction of the wind, but a westerly course prevails. These disturbances occur at intervals of three or four days in winter and with less frequency and strength in summer. At any point in the State, the wind may on occasion come from any point of the compass. The velocity varies widely and sometimes exceeds the speed of the fastest express train, but it seldom exhibits the type or the violence of the tornado. The southerly winds are most common in the extreme west and in the southeast part of the State.

In addition to the general wind movements, there are local winds, usually of moderate velocity. In the Long Island province a sea breeze is recognized in quiet weather. It is most prominent late in the day and may attain a velocity of ten to fifteen miles an hour. The land breeze is best developed in the latter part of the night. The sea breeze reaches inland about ten miles, and this may be augmented somewhat by the prevailing wind.

Adjacent to the Great Lakes a similar but less noticeable land and lake breeze is recognized. In the regions of steep topography, local breezes are developed by the flow of the cooled air down the slopes. Where narrow gorges connect elevated country with deep valleys, distinct breezes are observed in the evening that sometimes attain a velocity of eight or ten miles an hour in the mouth of these gorges. Such breezes are best developed in the plateau and mountain regions. These are drafts determined by unequal cooling of the atmosphere as soon as the sun gets low, coupled with their guidance by the form of the land surface. There is a less perceptible movement down broad slopes. These local movements have been observed in the valleys of southern New York often to be opposed entirely to the general wind movement 200 or 300 feet above the surface.[1] In the hill and mountain regions, the exposure of the crests and the higher slopes facing the prevailing storms to the vigor of the high winds, especially in the extreme temperatures of winter, makes a wide variation in conditions for living and for the growth of trees, and thereby affects agricultural development. Farms seek the sheltered positions for home sites. In addition the freedom from sweeping winds in such sheltered situations permits the snow to lay on the ground to prevent the destructive action of frequent freezing and thawing and protect winter crops from low temperature.

[1] Turner, E. T. "The Physical Geography of New York State," R. S. Tarr. Macmillan Co., 1902, p. 340.

The normal range in temperature that might be expected from the reach of the State across four degrees of latitude is four degrees Fahrenheit. The normal range in temperature that might be expected due to the range in elevation from sea level to 5,379 feet at the crest of Mount Marcy is approximately eighteen degrees Fahrenheit, or one for each 300 feet elevation. Long-time records show that the range in average temperature may exceed that, during the coldest months of winter when it reaches twenty degrees — ten degrees on either side of the normal. In summer the range of average temperature from year to year is ten to twelve degrees. Those regions which have the coolest mean climate are likely to have the largest range of extreme temperature.

A critical phase of temperature with reference to crop growth is the date of the last killing frost in spring and the first in fall, which mark the limits of the crop-growing period. The average length of the growing season determined by these limits is presented in Fig. 11. Less elevated regions have the greatest freedom from late and early frosts and, therefore, enjoy the longest growing season. Since each crop has an average minimum season in which it can mature, these weather conditions have a determining influence in the growth of particular crops. On the southern plateau, the maturity of the ordinary varieties of corn is very uncertain because of the shortness and irregularity of the seasons. In the mountain regions, severe frosts may occur any month in the year.

Proximity to large bodies of water also affects the

liability to untimely frost. Since water changes temperature much slower than land and the air in contact with the water is similarly affected, these land areas so situated as to receive the prevailing winds from over such bodies of water are correspondingly

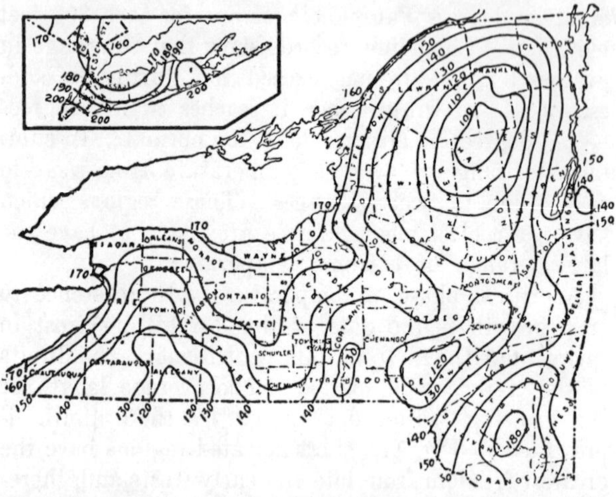

Fig. 11. Showing the average number of days between killing frosts in spring and fall which corresponds to the average length of the growing season.

tempered. This advantage is held by the Great Lakes plain. The water in Lake Ontario maintains an average temperature of ten to fifteen degrees warmer in winter and cooler in summer than the adjacent land areas. The success of growing fruit on the lake plains is largely due to the freedom from untimely frosts as well as the tempering influence of Lakes Erie and Ontario over which the prevailing wind comes into that region. This insures the better maturity of the wood and buds which increases their power of resistance. The greater uniformity of the winter weather keeps them dormant and resistant. The effect is especially noticeable on peaches.

The smaller inland lakes are considered to have a similar though smaller effect on the weather in their immediate vicinity and their ameliorating effect is said to be most noticeable on the eastern shore.

The weather of the Long Island and Lower Hudson Valley provinces is similarly modified. This has favored the development of the vegetable industry around New York City.

In the plateau province there may be a wide range in the length of season between the valleys and nearby hill regions. This difference is well illustrated by Ithaca and Perry City within twelve miles of each other in Tompkins County.

Liability to frost is also determined by topographic position in another way. As a result of the circulation of air in response to differences in temperature, the cold air settles into low pockets and deep flat valleys where frosts are frequent. On the highest points they are also likely to occur, due to rapid radiation and cooling at night. Consequently, the regions most free from frost are the intermediate elevations having a fair slope so that there is free circulation of air. The swamp and muck lands are generally liable to unseasonable frost because of these relations. The intermediate levels are the most desirable farm sites. Reference to the flow of water will often aid in reaching an estimate of the air circulation and liability to frost of a particular section.

Another element of temperature variation is the direction of slope. South and east slopes are warmest and west and north slopes are colder in the order named.

The mean annual rainfall for New York State ranges from 25 to more than 50 inches. The lowest rainfall occurs on the Great Lakes plain, in the central lakes region and in the upper part of the Champlain Valley.

The higher amount of rainfall is also in three regions, on the southwest slope of the Adirondack Mountains, on Long Island and over the heel of the State in the lower Hudson Valley. The Catskill Mountains and the eastern plateau section have a relatively high rainfall. The prevailing winds as they move east over the Great Lakes strike the rising country of the Adirondacks and by passing up this slope are lowered in temperature so that a part of their moisture must be precipitated. The same thing occurs to the winds that sweep up from the Atlantic and encounter the highlands in the lower part of the Hudson Valley and Catskills. On the other hand, the low rainfall in the Central Lakes region and the Champlain Valley occurs where the wind has passed over an elevation and its capacity for moisture is increased by its rising temperature. On the Great Lakes shore the wind from the lakes encounters the warming influence of the land with low elevation which increases its capacity for moisture and reduces the precipitation. The mean precipitation is shown in Fig. 12.

The precipitation is greater in summer than in winter, in many ways a propitious fact. It insures

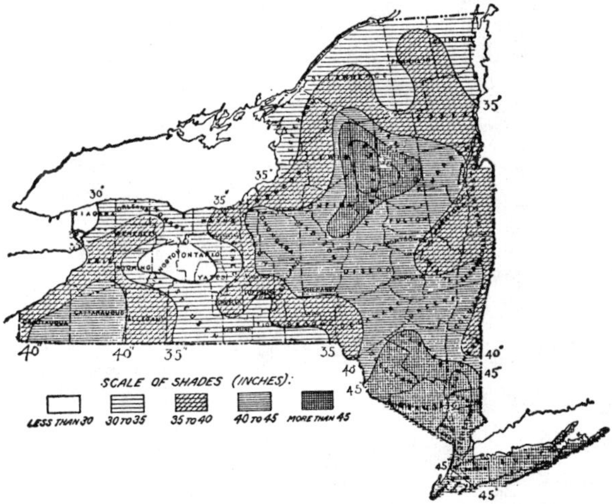

Fig. 12. Average annual precipitation in different parts of the State.

a better supply of moisture for crop growth, and by reducing the accumulation of snow in winter reduces the damage from floods due to melting snow, although it may sometimes expose winter crops to a little more severe climate and physical stress due to the lack of the snow covering.

The mean summer precipitation is shown in Fig. 13. The winter precipitation falls largely as snow. The regions of high precipitation are in the main those of low winter temperature. Long Island and the lower Hudson Valley are an exception to the rule.

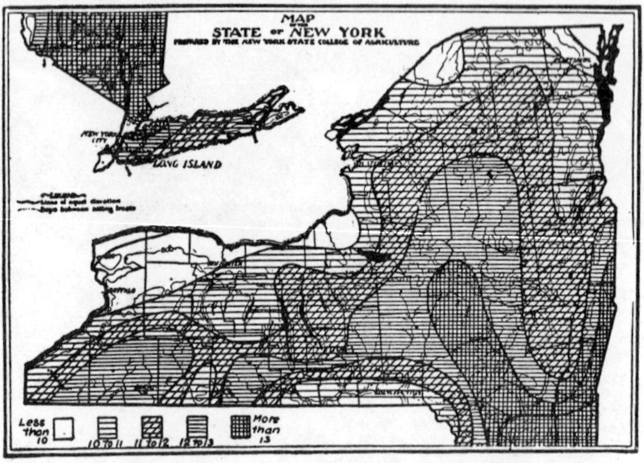

Fig. 13. Average amount of precipitation in inches in the three summer months of June, July, and August in different parts of the State.

Local storms and thundershowers develop most frequently on the western half of the southern plateau and on the western flank of the Adirondack Mountains. They travel eastward, a little more rapidly through the Mohawk Valley regions than over the uplands. The hour of the loudest thunder which is synonymous with the most violent electrical disturbance is about two or three o'clock in the afternoon at Buffalo and is progressively later toward the east until it occurs between twelve and one o'clock at night in the lower Hudson Valley region. The topographic hollow in the Finger Lakes region is most subject to secondary local storms.

The proportion of cloudiness which is the obverse of clear weather varies as a mean over the State from a little less than 50 per cent to over 65 per cent. The higher proportion of cloudiness is not directly associated with the precipitation. Southeastern New York has the largest amount of clear weather since the moisture-ladened winds from the ocean tend to drift eastward. The winter is more cloudy than the summer, especially in the Great Lakes region where the range is from 80 or 85 per cent in December and January to about 45 per cent in July and August.

The relative humidity is another factor in the weather, of considerable importance both to crop production and personal comfort. High humidity not only makes extremes of temperature more unpleasant but it promotes the growth of fungous organisms, including many disease-producing forms. The annual humidity at different points ranges from 70 to 79 per cent in the lowland areas, to 60 in the upland regions. It is highest in the vicinity of Buffalo and New York City. In general it is higher in winter than in summer.

Sleet storms are most common a little way inland from the shore of the Great Lakes. Hail is most common in the Central Lakes region and on the northern rim of the southern plateau where also local storms most frequently occur.

Disastrous floods occur in all parts of the State on occasion and are determined somewhat by the condition of the drainage system. Abnormal rainfall is most common in the Central Lakes and southern plateau region in association with the prevalence of local storms. In such cases a rainfall of several inches may fall in a few hours and constitute what is known as a "cloudburst." Such downpours are very destructive as a result of erosion and flooding.

The distribution of crops in New York State is determined in considerable measure by these local differences in climate. The fruit regions have a long growing season and a stability of winter climate that protect against winter-injury. Corn, beans, cabbage and vegetables thrive best in the lowland regions of high summer temperature and high humidity. This localizes their extensive production very largely to the central part of the Great Lakes province and the Long Island and lower Hudson divisions. Potatoes are grown most extensively under the cool climate of Long Island, northern New York and the west central plateau region. Different factors combine in these regions to produce conditions favorable to particular crops. Grapes are grown under the influence of the dry sunny climate of the Erie Lake plain and the Central Lakes region, combined with immunity from frost, part of which is secured by elevation and the careful selection of the slope. For the forage and vegetable crops, one of the most determinate factors in success is the summer rainfall and curves showing the total rainfall in the summer season and the yield of hay and corn correspond very closely. Departures from the normal weather elements are large and in the case of rainfall raise the question of the practicability of irrigation. This depends on the actual frequency and extent of variation of the rainfall, the nature and value of the crops and the cost of applying water. In general, irrigation will be practicable only on crops of high acre value and then only on the coarser soils of low moisture capacity.

CHAPTER II

HISTORY OF AGRICULTURE IN NEW YORK

THE first white men came into New York in 1609, two years after the settlement at Jamestown, Virginia, and eleven years before the Puritans landed at Plymouth, Massachusetts. In July of that year Champlain entered the lake region, that now bears his name, from the north, and Captain Hendrick Hudson discovered the Hudson River and spent the month of September on it. He sailed up the river to the head of tide-water at what is now Albany. With his crew he interchanged courtesies with the Indians and later bought of them fruits, maize and other products of the soil.

The permanent settlement of New York began in 1623. In that year Peter Minuet acquired from the Indians their claim to Manhattan ("Manna hatta") Island for twenty-four dollars. Farms, which were called boweries, were laid out on that island and on adjacent parts of Long Island. To the region the Dutch gave the name New Amsterdam. In succeeding years, under supervision of the Dutch West Indies Company, settlements of Dutch were planted along the Hudson River ás far north as Albany or Fort Orange and among these was a considerable number of Walloons or Belgian French.

In considering the development of the settlement, it is important to keep clearly in mind the physical nature of the country and the more accessible lines of travel, which were the waterways and the low passes between the hills. The new settlement was peculiarly situated in that respect, a fact that was more vital to its later development than to its early expansion. The wonderful harbor, the navigable tide-water as far as Albany together with the low pass into the interior of New York State, and also the great middle section of the United States, through the Mohawk Valley and the Great Lakes region, have coöperated to make this route the main gateway to the middle country and to maintain an organic connection between the peoples and customs in this eastern region and those in the more westerly districts.

AGRICULTURE OF THE INDIANS

The territory of central New York from the eastern part of the Mohawk Valley to west of the Genesee River was the home and the stronghold of the Five and later the Six Nations of Indians to whom the French gave the name Iroquois. They called themselves Hodenosounce and in some cases Onguie-Honwe, the latter meaning "the men or people of the Long House." This term was descriptive of the type of their dwelling-houses which were long frame structures of poles and bark, built in compartments, and in which lived several families of the same clan. (See Plate II.)

The geographic situation of this Indian confederacy was peculiarly strategic. It lay at the headwaters of great rivers that flow to all points of the compass. It is the setting for an empire and New York seeks to live up to that standard of outlook, as is indicated by its popular synonym "The Empire State." In view of the limited means of travel, the situation of the Iroquois gave them a distinct advantage in conflict with the neighboring tribes, all of which tended to strengthen their confederacy and to build up a high state of social and industrial development. They probably acquired many useful ideas and customs from the tribes with which they had contact.

In form of government, in the structure of their houses and villages, and in their agricultural practices, the Iroquois had reached a high state of development. They were composed originally of five tribes — the Mohawks, Oneidas, Onondagas, Cayugas, and the Senecas, situated in regions of which their names are still indicative. Later, the Tuscaroras of North Carolina were conquered and adopted into the Confederacy, and occupied the region adjacent to the Niagara River. The tribal councils of sachems and chiefs were assembled in inter-tribal conference at the council house or castle in the Onondaga Valley. They were bound together by recognized principles of coöperation and the rudiments of democratic government in which female suffrage was recognized. It is said that the early colonial assemblies in New York that led ultimately to the Articles of Confedera-

Plate II. Model of Iroquois Indian house, stockade and family settlement.

tion and the Constitution of the United States were patterned somewhat after the Indian councils. The Indian Confederacy was claimed to have existed for six generations. The total Iroquois Indian population is said never to have numbered more than fourteen thousand after the advent of the white man. In addition, there were the Erie and other tribes in the western counties.

The Iroquois had numerous villages protected by stockades of poles driven into the ground and equipped with defensive platforms. Adjacent to the villages were cleared and cultivated fields sometimes reaching one or two hundred acres in extent, usually on low land along rivers or lakes. Notable among these were fields at Indian Castle, east of Little Falls on the flats of the Mohawk, Onondaga Castle, south of Syracuse, Seneca Castle near Geneva, at Honeoye Falls, and a very large field in the Genesee Valley between Geneseo (Big Tree) and the present site of Mt. Morris. Other important settlements were near Elmira, Binghamton and at Peach Orchard Point, now Aurora.

Maize and beans, pumpkins and squash seem to have been the main crops. General Sullivan's raid in 1779, by which the strength of the Indians in central New York was broken, is reported to have destroyed great stores of corn and other crops at the different settlements aggregating more than 200,000 bushels. The maize included flint, sweet and pop varieties. In addition, peas, turnips and other vegetables and tobacco were grown. Winter stores were preserved in chests and in deep pits. Vegetables were dried and meats were smoked, but the act of preserving with salt was not known. Many native fruits, including the wild grape, were used. With the advent of the white man, the Indians were brought into touch with the European fruits, the apple, peach, cherry and pear. They were particularly attracted by the apple and developed extensive plantings. They had an orchard on Stockbridge Hill in Madison County, and early observers report that in the Seneca Lake region and in the Genesee Valley there were thousands of apple trees, some of which near Geneva were still standing as late as the opening of the twentieth century. Peach Orchard Point, or Aurora, is suggestive of the growth of peaches there and trees of that fruit were scattered through the later Indian settlement. From certain reports the Indians seem early to have adopted some live-stock, but the abundance of wild game and fish made domestic animals scarcely necessary to their sustenance.

EARLY SETTLEMENT BY THE WHITE MEN

The spread of the early population was slow. Fifty years after the first settlement, Schenectady was the most remote organized outpost. During the eighteenth century, settlements spread up the Mohawk Valley, and as late as 1790 in all western New York beyond Cayuga Lake, there were only about one thousand white persons. To encourage settlement and develop trade, the Dutch West Indies Company arranged for grants of large tracts of land to persons who would bring into the new country fifty adult settlers within four years. These grants were to have a frontage of sixteen miles on the Hudson River, eight on either shore, and with indefinite boundaries inland. This was the basis of the Patroon settlements by which the owners of these grants were known. The Patroon was obliged to divide his land into farms and to aid in their equipment, with buildings, tools, seeds, plants and stock. Rent was in stock and produce, usually fifty bushels of wheat a farm. Tenancy was for ten years, during which the tenant was securely bound to the Patroon who was officer and court. In addition to the first right to purchase the tenant's products, he controlled the rights to establish mills and to fish and hunt.

The largest and most noteworthy grant of this sort was to Killan Van Rensselaer who acquired 700,000 acres extending from Albany, which he founded in 1630, twenty-five miles south along the river and forty-five miles inland. In 1684 after New Amsterdam had become permanently English and changed its name to New York, Chancellor Livingston acquired a tract of 160,240 acres further south in the Hudson Valley in what is now lower Columbia and northern Dutchess counties. By the English these grants were made from time to time but the principle of tenancy rather than actual ownership by the operators of the land was firmly established and led to important agrarian movements that are still visible in the political structure of the State.

In 1685 the Huguenots or French Protestants came

into Ulster County, having already a foothold in Westchester County into which Puritans from New England and Quakers from Long Island and Rhode Island had filtered. About the beginning of the eighteenth century the West German Protestants, or Palatines, were very much unsettled in their home country and through the influence of Governor Hunter of New York, 3,000 came to the colony, 2,277 of them were distributed through the settlements up the Hudson Valley. Many were located on Livingston Manor and got out tar and resin. Within a year there was a strike and some of these settlers were induced by the Indians to go into the Mohawk Valley where they located at Stone Arabia and Palatine Bridge in Montgomery County.

To a considerable extent the pioneer settlements throughout the early American colonies were a Protestant search for religious liberty. Puritan, Palatine, Huguenot, Quaker and Adventist, Methodist, Baptist and Episcopal with many other sects have formed strong chains of settlements that spread throughout the State from their earlier foothold in this east country.

The introduction of Dutch immigrants was not active after the acquisition of New York by the English. Then the latter, largely from New England, spread westward in nearly straight lines modified, of course, by the easier lines of travel. They dominated the settlement of all western New York, probably in large part because much of that region was claimed by Massachusetts, which retained preëmption rights even after the relinquishment of its political claim about 1790. Welsh came into lower Oneida County, Scotch and Irish into Otsego and some Scotch into Genesee, French filtered into northern New York and the St. Lawrence Valley from lower Canada. Some Germans settled in western New York from Pennsylvania. The different nationalities came somewhat in successive waves. After the Revolutionary War, many German and English soldiers remained as settlers, and Yankees from New England took up the confiscated Tory lands, such as those of Sir William Johnson in the Mohawk country. After Sullivan's expedition into western New York in 1779, many of his soldiers returned to that region as settlers. After the early stocks came the Irish in the early part of the nineteenth century, many Germans in the middle, and Poles, Hungarians and Italians in the latter part. All these have been distributed widely wherever large public works have been constructed, and they have remained in those regions to enter more permanent industries, including agriculture. The early settlements of different nationalities and religions soon became so much intermingled that there is now very little evidence of the persistence or domination of the peculiar agricultural customs of any one group. Perhaps the inclination of the Germans to establish small farms is one such custom. (See Fig. 14.)

On the whole, the various nationalities and the different religious sects have blended their customs and have developed a general tolerance quite characteristic of the New World.

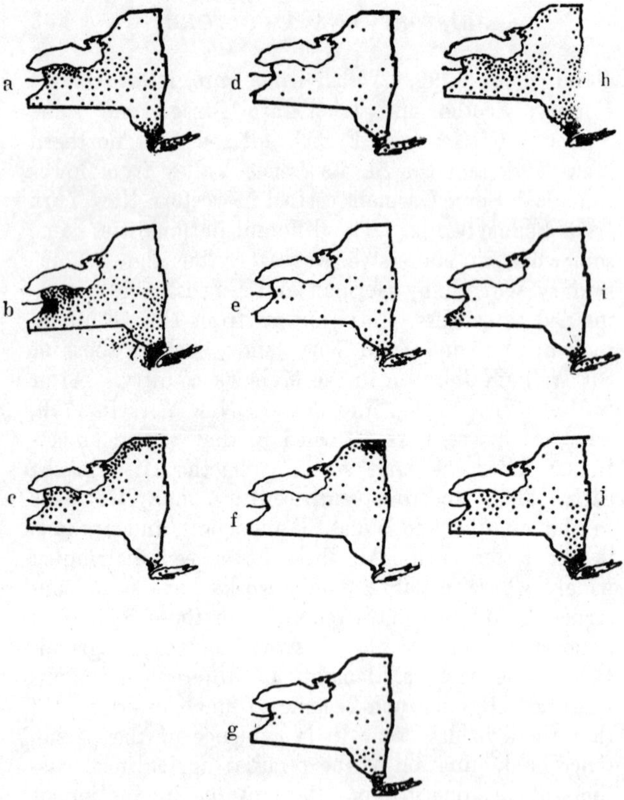

Fig. 14. Numbers and distribution of rural population of foreign stock by nationalities in 1910. One dot equals 500 persons: a. English; b. German; c. Canadian (not French); d. Scotch; e. Austrian; f. French; g. Russian; h. Irish; i. Norwegian; j. Italian.

EARLY ROUTES AND MEANS OF TRAVEL

Of the early routes of travel, the Hudson and the Mohawk valleys have been mentioned. At the present site of Rome was the portage to Wood Creek from which Oneida Lake and the Seneca and Oswego rivers were accessible and led to the Finger Lakes and Lake Ontario. The rivers, lakes and the other parallel valleys opened the way to the middle hill region. From Utica a line of travel led up Black River to the Jefferson County region. The Lake George and Lake Champlain valleys opened up the country to the north. In lower New York, low passes led through the Catskills to the headwaters of the Delaware and Susquehanna rivers on which it was possible to float down stream into the new country. The boats were also pushed up these rivers by poles. From the Susquehanna its westerly tributaries could be followed into the hill regions of lower western New York, past Elmira and Hornell to the headwaters of the Genesee River. In western New York, the Genesee River was followed up to the present region of Belmont or beyond, from which it was possible to push a short distance overland to the Allegheny River at Olean.

Many of these routes were traversed later by a canal and still later by railroads. The first canal in America is said to have been constructed in Orange County in 1750. The great movement for the construction of canals began about 1790 and was pushed during the next twenty years both by the elder and the younger Governors Clinton and culminated in the construction of the Erie Canal which was opened in 1825. From the main line of the Erie Canal a number of branches were constructed in the subsequent fifteen years. Since New York State has always been prëeminently agricultural, the opening of these lines of travel and ready means of moving heavy freight were a great stimulus to its development by affording an outlet for its products. Previously the market for farm products had been essentially local. The strength of this statement will be recognized by considering that the only other method of movement of farm products was by wagon or the movement of stock on foot. Wheat was hauled from the Finger Lakes region to Albany. In 1791, General Joshua Whitney drove cattle from Binghamton to market in Philadelphia and returned by flat (Durham) boat and wagon with merchandise. The most important period of development in New York was the first half of the nineteenth century and especially following the construction of the canals and the first railways. Shortly after the opening of the Erie Canal, the Hudson Valley farmers complained that Genesee wheat was flooding their market. Northern New York was effectively settled about the same time as western New York and as a part of the same general movement.

Antedating canals were the trails and wagon roads. A post road was early established along the Hudson from New York to Albany. A good road was extended up to Clinton County. The Newburg and Cohocton turnpike led to the Delaware River and central Sullivan County. A road was early developed along the Mohawk and at Utica left the river and followed a course by Wampsville and Syracuse to Auburn, and thence to Geneva and Canandaigua. It was pushed through to Buffalo with branches to Geneseo and a north fork west of the Genesee River leading to Lewiston on the Niagara. This road was opened to Auburn in 1789 and to the Genesee in 1791. At that time the western half was known as the Genesee road and as it developed sections passed under different names, such as Seneca Turnpike. From the region of Utica a northern branch swung around by Camden into Jefferson County, and as a result of a special improvement was known as the Plank road. Many of these early roads were improved through the swamps by means of logs placed crosswise and more or less covered with earth, known as Corduroy. These were more substantial than comfortable to ride over, especially in the common springless wagon of the region. Many of these old trunk line roads have now been linked up in the state improved highways system over which automobiles glide with great speed where once oxen drew carts that made slow, rumbling and jolted progress. Linked with the early development of the canal was the introduction of Fulton's steamboat, the *Clermont,* which on August 7, 1807, traversed the Hudson at the "remarkable speed" of a little more than five miles an hour.

The Erie Canal, which was originally forty feet

wide and three and one-half feet deep, and carried boats of seventy-five tons' capacity, was enlarged between 1847 and 1852 to a width of seventy-five feet and a depth of seven feet carrying one hundred fifty-four tons, and later was enlarged to accommodate two hundred twelve ton boats. As a result of the larger capacity, and probably also due to competition with the railroads in the later periods, the freight rates were successively reduced from $1.12 in 1839 to twelve cents in 1882 for carrying a ton from Buffalo to Albany. Between 1903 and 1920, the Erie Canal was enlarged with some changes in course to the form of the Barge Canal. It is twelve feet deep and will accommodate boats carrying three thousand tons which will be moved by steam tugs instead of by horses.

The construction of railroads followed close on the opening of the Erie Canal. By 1832 the railroad had reached Rochester and was operated by steam in 1837. The first line reaching across the State was the Erie through the southern part and the first train from New York reached Dunkirk in 1851. Next came the New York Central which, because of the low elevation of its bed along the line of the Erie Canal and the Hudson River, when taken in connection with the large agricultural production along its route, has come to be one of the heaviest lines of travel on the continent.

Nearly all the existing lines of railroad were constructed by 1880. After 1895 the electric railroads were developed along the lines of the more intensive local travel, often as branches to accommodate local traffic. Now the automobile truck and bus lines are cutting into the business of the electric and also of the steam lines for local business, and by reason of their elasticity are able to reach out into strictly rural districts that were formerly very much handicapped because of lack of transportation facilities. All these means of travel have stimulated agricultural production but when coupled with their application to the remainder of the country have brought trying industrial problems through the wider range of competition.

TERRITORIAL DEVELOPMENT

The territorial expansion and organization of the State are somewhat intricate to trace in any complete way because the boundaries of the counties were changed from time to time and several counties were often made from a single earlier one. Much of the original ownership of the land by the white men rested on large grants by European crowns, or purchases from the Indians, and on claims of the more eastern colonies.

In the Hudson Valley the Dutch Patroons and the English Manor estates occupied large tracts and developed an agricultural tenantry. But in a country with so much cheap land, such a system could not be very exacting as to rents. Long-time leases were given and in the end the landlord was so far separated from his land that by the end of the period, when he came to claim his due, the tenant who had in the meantime made extensive improvement was likely to resent the intrusion of such claims by the nominal owner. This was the feudal tenure of the Patroons, the practice of which between individuals was prohibited by acts of the legislature of 1789. But that act did not abolish the difficulties already started. They were aggravated wherever there were large estates in the ownership of absentee landlords. Conflicts of this sort fermented during the first part of the nineteenth century in the Hudson Valley region and later spread to the lands owned and disposed of on contract by the Holland Land Company in western New York. These culminated in the anti-rent wars of 1836 to 1845 during which there were armed conflicts in the east resulting in bloodshed, such as those at Grafton, Rensselaer County and at Reidsville in Albany County in 1839, and the burning of the office and records of the Holland Land Company at Mayville, Chautauqua County in 1836. In 1846, the legislature set a limit of ten years to leases and abolished all feudal tenures. This principle is now established in the Constitution of the State.

The ownership of the farm land of the State is now almost exclusively in small areas suitable for individual farms or at most a very few farms, except in the Genesee Valley where many thousand acres are owned by the Wadsworth families, title to which has come by inheritance from the pioneer acquisition. Even these areas are divided into relatively small farms and are operated under the ordinary lease system. In this region is the nearest approach afforded by the State to the English manorial and social system.

The Colonial assembly of 1683 established twelve counties: New York, Richmond, Kings, Queens, Suffolk, Westchester, Dutchess, Orange, Ulster and Albany, and these included the present area of Rockland, Putnam and part of Columbia, and at the west of Albany all the Mohawk Valley country as far as any jurisdiction was claimed.

Tryon County was set off from Albany in 1784 and embraced all central and western New York and out of its territory more than a score of counties were created directly and many others ultimately. Its last remnant became Montgomery County. Herkimer County, named after General Herkimer, the first chairman of the home defense committee of Tryon County, was set off from the latter, and from it Onondaga and other counties were formed in 1788. This region was largely comprised within the Massachusetts Military tract of 600,000 acres lying between the Oswego and Chenango rivers, acquired from the Indians by the treaty of Fort Stanwix in 1768. From it the counties of Onondaga (1817), Cayuga (1817), Seneca (1816), Cortland (1808) and parts of Tompkins (1817) and Oswego were carved. Madison was created in 1806 and Oneida in 1816; Broome County was formed from Tioga in 1806. In 1779 all western New York beyond Seneca Lake was included in the one county of Ontario, and at that time had a total recorded population of only 1084 persons.

The vast rich country was involved with the Military tract in the claims of the State of Massachusetts, under grant of the King of England. Its claim was adjusted with New York on December 16, 1786, by which Massachusetts retained the preëmption right to settlement while all political authority was reserved to New York. The eastern boundary of this latter territory was roughly a line north and south through Seneca Lake (due north from the 82nd milestone on the Pennsylvania line to Lake Ontario) and was thereafter known as the Preëmption line. The preëmption rights of Massachusetts were acquired by Nathaniel Gorham and Oliver Phelps in 1788 for 300,000 pounds consolidated securities and later they purchased the claims of the Indians to some two and a quarter million acres scattered through the western part of the State. The main part of the tract was east of the Genesee River and was known as the Genesee tract. As a result of financial stringency, the titles of Phelps and Gorham were acquired in 1790 by Robert Morris and by him disposed of to a Dutch company made up of Sir William Poultney, William Hornby and Patrick Calquahan, and known as the Holland Land Company. This concern and its successors surveyed the area and opened land offices at Canandaigua, Batavia, Mayville and other points, and gradually during the next sixty or seventy years disposed of its holdings.

Prices of land in the early days in these large areas as well as in small farm tracts were, of course, low, as the supply of land was almost unlimited. Then, as later, there was land grabbing and speculation. Individuals acquired large tracts in various ways at very small cost, both before and long after the Revo-

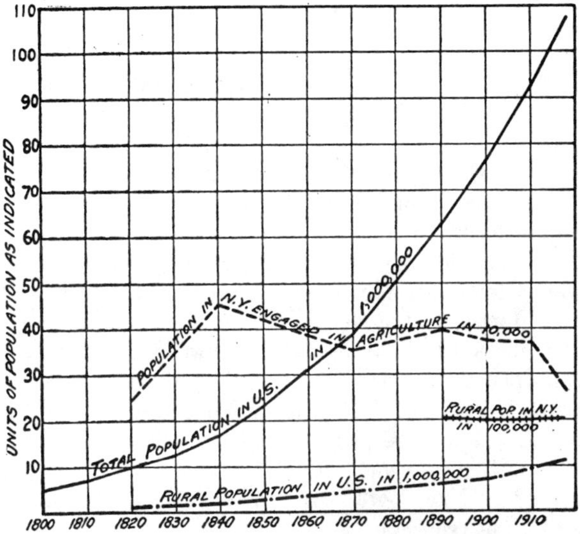

Fig. 15. Graphs showing by decades the total and the rural population of United States, the rural population, and the population engaged in agriculture in New York.

lutionary War. In 1791, 5,542,173 acres of land were sold in western New York for a total of $1,030,-433, of which Alexander McCombs is reported to have secured 3,635,202 acres, some of it as low as eight pence an acre. It sold to settlers for one to six dollars an acre.

Ontario was broken up into a total of fifteen counties and one of the earlier of these was Genesee which in 1802 included nearly all the land west of the Genesee River. Nearly all of the counties in that region had acquired approximately their present boundaries by 1824.

Nearly all of New York State was originally heavily timbered. The growth was especially heavy and diversified in the calcareous soil regions of the middle part of the State. Hardwood made up nearly all of the tree flora. Further south on the hills were much heavy white pine and several varieties of Coniferae dominated the mountain regions. There were a few prairie tracts mostly in river bottoms. At Mt. Morris, in the Genesee River flats, was a prairie of 6000 acres. Through the hardwood country were oak openings or tracts, sparsely covered by trees and brush. On Long Island around Hempstead and Jamaica was a large prairie tract on well drained soil, a condition unusual outside of the western prairie country. New York was both a well watered and a well timbered country.

To clear the land the timber was cut and burned, and in many settlements the first money was secured from the sale of ashes or from the potash and pearl ash salts derived from them. It must be remembered that in that day potassium nitrate was the basis of gunpowder, the only available explosive, and the potash secured by leaching the ashes was necessary to its manufacture. In some settlements, for example Clinton County, charcoal was produced.

Tanneries were common everywhere, and grist and saw mills sprang up wherever a little water power could be developed and became relatively numerous soon after settlement in every district.

As soon as there was any market and a means of transportation, lumbering became the leading industry. The opening of the Erie Canal and its tributaries was a great stimulus to the business, and in 1832 Albany was the largest lumber market on the continent. At that date the lower Hudson Valley was relatively an old settled region in need of that product. The first cash crop was generally wheat, the production of which in western New York was also given a tremendous impetus by the opening of the canal. Rochester and Oswego became great flour and barrel manufacturing centers, the former being known as the Flour City. Genesee Valley wheat was the standard in the market. Later, when the production of wheat declined and Rochester ceased to be the great flour center of the North, it was happily able consistently to change its popular name to the Flower City.

CROP DEVELOPMENT

The relative disadvantage in character of soil and topography in some parts of New York to that in the Middle West was greatly accentuated by the coincident development of modern farm machinery especially adapted to broad level acres and by the rapid transportation facilities afforded by the steam train. After 1875, land values shrank rapidly and only the best farms could be maintained in active tillage in the open competition. In each decade in the last third of the nineteenth century, the cultivated area in New York was less than in the preceding one. Stock, espe-

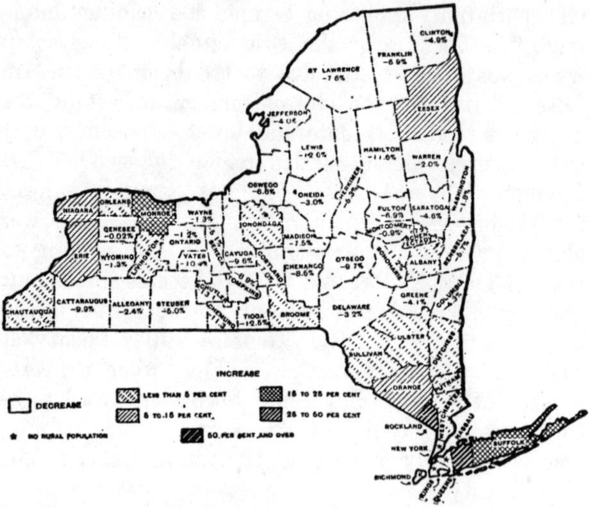

Fig. 16. Changes in rural population from 1900 to 1910.

cially sheep and hogs and beef cattle, was reduced in numbers together with the reduction in acreage of tilled crops. Not until the present time has the shrinkage in acreage of crops and of certain types of live-stock reached its base level from which it may now be expected to ascend.

The largest proportion of the total land area of the

state in farms was about 23,750,000 acres or 78 per cent in 1880. The area in improved land was also largest at that time and was 17,700,000 acres or 58 per cent of the land area. Figs. 15 and 16 show the rural population of the State.

The largest total acreage in hay and forage was in 1870 when it was 5,600,000. The State has always had a large area in forest and timber land and besides the 22 per cent not accounted for in farms in 1880, or an even larger percentage in other years, a considerable proportion of the farm land has always been kept in timber so that never less than one-third of the State has been so occupied, and a further increase is likely to occur. Grass and pasture have always occupied a large area, partly because of the land that was too rough and stony to be well adapted to other purposes, partly due to the fact that the climate of the State is well suited to these crops, to the large proportion of springy wet soil, and to the relatively profitable demand for hay and pasture. The area in cultivated crops other than hay has always been low in proportion to the total, and since 1844 has swung very close to 4,000,000 acres, or about 13 per cent of the total. It has been from 20 to 25 per cent of the total improved area in farms and 50 to 60 per cent of the total land in specified cultivated crops. The largest total acreage in crops other than hay and fruits in this period was in 1854 when it was nearly 4,200,000. It was low during the Civil War period and high again between 1875 and 1880. Since 1909, it has been near 3,500,000 acres.

The area in fruits cannot accurately be determined before 1909 when it was 500,000 acres, and this had increased to 694,000 in 1917. Before 1890, the proportion of commercial orchards was small, most of the main area being for home use. The increase since that date has been mostly of the commercial or shipping type. Even yet, the proportion of the total acreage of fruit that is actively put on the market is relatively small and probably does not exceed one-third of the crop in a normal season.

The land area is still far from well developed. The productive capacity of the land has decreased on particular farms and in special regions. The problem of maintaining fertility has come to the foreground. But from a consideration of the average yield of all crops over a period of forty-seven years, from 1866 to 1912, it is difficult to prove a decrease in yields. Taking the yield in 1866 as 100, there have been nineteen years when the average yield was equal to or greater than this figure and twenty-eight when it was less. From 1879 to 1896 inclusive, there was not a year when the yield was above the base figure. This was the period of agricultural depression and readjustment in the East. In the period from 1897 to 1912, the yield has been above that base figure eleven times. In these figures is expressed the more careful adjustment to areas adapted to the cultivation of the main crops of the State in order to meet the outside competition.

DEVELOPMENT OF UNDER-DRAINAGE

In the practice of modern under-drainage by means of tile, New York was a pioneer. Drainage as an art accomplished by the use of open ditches and of stone, brush, poles, and other crude means is very old and is recorded by Cato and other ancient writers. But drainage by means of tile or short lengths of clay pipes is relatively new and probably does not date back more than 250 years to the convent gardens in Mauberg, France. In England it is not much over 150 years old.

The honor of having first systematically tile-drained a farm in America rests with John Johnston, a Scotchman, who came to America in 1821 and acquired a farm two miles southeast of Geneva, on the east shore of Seneca Lake. His farm is in part a strong heavy calcareous loam and in part a rather heavy calcareous clay. For the first fifteen years, however, he could not grow profitable crops. Then he remembered the "pottery" he had seen buried in the land of his native country. In 1835–1837, he imported some of these clay pipes from England and began the systematic drainage of his farm about four rods apart. He found it so profitable that he continued until his entire three hundred acres were all well drained in this manner. Others followed his example and a period of active tile under-drainage began. Johnston was associated with John Delafield in importing the first machine for making tile, a Scragg pattern. In 1851, Johnston was awarded a prize by the New York State Agricultural Society for an essay on his drainage operations, and in the following year a set of exquisite gold and silver pieces was presented to him in recognition of his service to agriculture in introducing tile drainage, by a group of public spirited men, among whom are a number of prominent names. Whether the Johnston farm was actually the first in America on which clay tile were used is an open question. South of the Mason-Dixon line, tile were not systematically laid until about 1875 and this was at Charleston, South Carolina. W. C. Hinson, a planter on James Island, then began using tile. It is reported that one of the English consuls stationed at a Virginia port used tile long before that date. In any event, Johnston's was the first work the results of which have been carried through to the present day, for those drains are still operating and in good form and the farm, under the progressive management of the present owner, continues to produce yields of crops far above the average of the State.

The first sections of farm drain tile carried west of the Mississippi River came from this same vicinity of Geneva. They were carried by another man who has had very large influence on agriculture in New York and also in the country at large. Isaac Phillips Roberts, first dean of agriculture in Cornell University, was born on a farm at East Varick on the west shore of Cayuga Lake. He moved to Iowa and engaged in farming. In 1865 he carried some lengths of tile back to Iowa in connection with a visit at his old home in New York. He preached the gospel of under-drainage in the new country then and later in his capacity as a professor of agriculture in the new State College of Agriculture at Ames, Iowa.

During the depression in agriculture in the last quarter of the nineteenth century, New York fell behind the Middle West in the practice of tile drainage and is just now actively resuming the art.

THE RISE OF AGRICULTURAL INSTITUTIONS

The institutions and educational movements that have so large a part in the agricultural progress and prosperity of the State have been of slow growth. The antecedents of these were the societies for the promotion of agriculture, both local and state. In February, 1791, The New York State Society for the Promotion of Agriculture was organized, and at once became a potent force in the discussion of agricultural matters. The movement by which exhibits of agricultural products were held, with prizes for the best, had its rise between 1800 and 1810, in which latter year Elkanah Watson conducted a cattle show at Pittsfield, Massachusetts, that led to the organization of the Berkshire Agricultural Society. In 1817 an agricultural fair was held at Red Hook, Dutchess County, and in 1818 a similar one at Auburn. In January of that same year, Governor De Witt Clinton made a notable address to the state legislature in which he set forth the need of more attention to the practical and educational needs of agriculture and recommended the formation of a Board of Agriculture. In the following year, such a board was constituted with an appropriation of $10,000, to be distributed among the several local secretaries for aid in agricultural affairs. This Board of Agriculture with the Society for the Promotion of Agriculture were merged in 1832 into the incorporated New York State Agricultural Society, which took over and continued until a very recent date the functions of the Board, under which an annual volume or repository of agricultural information was published. This annual volume was for many years the chief source of agricultural information and inspiration. That Society may be credited with the large amount of attention given to soils and fundamental agricultural facts in the five-volume report on the Natural History Survey of the State published in 1844. About 1849, the Society inaugurated the examination of soils, seeds, fertilizers and food stuffs, which movement became the lineal ancestor of the present division of Farms of the Council of Farms and Markets, which was organized as the office of the Dairy Commission in 1884 and became the Department of Agriculture in 1893.

The agricultural fair movement had its largest development in the period of 1850 to 1870. In 1857 there were ninety-seven agricultural societies or fair associations in the State. The early policy of state contributions to the premium list of these fairs has been continued to the present date and for many years was paid from the receipts from licensed gambling on horse races. A state fair was first held in Syracuse in 1841 and for many years it convened in different parts of the State. In 1890 it was permanently located at Syracuse by the State Agricultural Society and one hundred acres of land on the present site were donated by that city. This area has been increased to 145½ acres. At that time the development of the present commodious grounds and buildings was begun. In 1900 the enterprise was taken over by the State and was placed under a commission of eleven men. In 1909 this number was reduced to seven, with the Lieutenant-Governor and Commissioner of Agriculture ex-officio.

From the time of the notable addresses to the legislature of Governor De Witt Clinton in 1818, in favor of agriculture, and the writings of his cousin Simeon De Witt, the founder of Ithaca, both nephews of the first governor of the State, Robert Clinton, who was also a strong advocate of the promotion of agriculture, various movements were started for the organization of a school or college of agriculture. Rensselaer Polytechnic Institute, established at Troy in 1824, included agriculture among its interests. After long and persistent agitation in the New York State Agricultural Society, a charter for an agricultural college at Fayette, Seneca County, was granted in 1853, largely through the efforts of John Delafield. As a result of his death in the same year, the movement failed and was supplanted in 1856–1857 by the organization of an agricultural college at Ovid, in the same county, in connection with Ovid Academy, where buildings were erected and the doors opened to students in agriculture in 1860. But this was not a permanent start. There was in that first faculty William H. Brewer, a native of Ithaca, who for forty years, from 1864 to 1904, was professor of agriculture in the Sheffield Scientific School associated with Yale University. Another man who took an important part in the launching of agricultural colleges appears in the records at this time. This was the Rev. Amos Brown who had been principal of the Ovid Academy and the School of Agriculture and was transferred to the presidency of the People's College at Montour Falls, in Schuyler County. This institution was started in that year as an outgrowth of a demand among mechanics for vocational education in the mechanical trades, and agriculture was included among its interests. The significant fact is that Amos Brown was personal advisor of Senator Justin Morrill, who was instrumental in securing the passage of the Morrill Act of 1862, by which public lands were given to each of the states of the Union, from the proceeds of the sale of which colleges of agriculture and mechanic arts were to be established. That act is the basis of such institutions in all the states, a number of which have grown into universities of first rank. The first assignment of the funds accruing from the land scrip credited to New York was to the institution at Montour Falls, of which Brown was president. Owing to failure of that institution to comply with the conditions of the assignment, the outgrowth was the transfer of this grant and the founding of Cornell University at Ithaca. From its opening in 1868, instruction in agriculture and the mechanic arts has been included, together with the old established arts subjects and the newer instruction in the sciences. The department of agriculture in that institution was taken over by the State in 1904 and chartered as the New York State College of Agriculture. The period between those two dates, 1868 to 1903, marks the full crystallization of the idea of collegiate instruction in agriculture based on careful scientific investigations as represented by the agricultural experiment stations, which were an outgrowth of the colleges. In 1879, the Cornell University Agricultural Experiment Station was organized by the faculty in agriculture, and in the same year the students of agriculture in the same university inaugurated a movement which resulted in the founding, in 1882, of the New York State Agricultural Experiment Station at Geneva, which was the second institution of its kind in the country, and antedated by five years the foundation of such institutions in all the states by the federal act of 1887. As early as 1876, a private agricultural experiment station was established at Houghton Farm, in Orange County. It continued until the death of its founder, Lawson Valentine, in 1891.

Isaac P. Roberts, the first permanent professor of agriculture in Cornell University, and for many years the dean of agricultural teachers in America, was the director of the Agricultural Experiment Station at Ithaca. His period of service, from 1874 to 1903, as head of the agricultural work at Ithaca, covered the pioneer period in agricultural education. That of his successor, Liberty Hyde Bailey, represented the attainment of agricultural education to equal rank with other academic courses, and the agricultural writings and publications of the latter have been a leading force to popularize agricultural practice and rural life, and to give it a standing in the literature of the day. The Station at Geneva, which was one of the first in America that was well organized and took up a carefully considered program of experimentation and investigation, had as its first director E. L. Sturtevant, who was succeeded in 1887 by Peter Collier, who in turn was followed by W. H. Jordan in 1895.

The beginnings of the farmers' institute movement are hazy and are lost in the miscellaneous lectures on agriculture that began at a very early date. The organized movement commenced with a series of meetings of practical farmers with the agricultural teachers at Ithaca in 1886. Out of this grew the scheduled series of lectures on agriculture by successful farmers and by members of the staff of the agricultural institutions, under state financed subsidy. The institutes were for many years administered by the State Department of Agriculture. On the death of the last director, Edward Van Alstyne, whose sturdy teachings reached beyond the bounds of technical farm problems into the larger social and spiritual domains of rural life, most of this work was transferred to the supervision of the State College of Agriculture at Ithaca. This occurred in 1917 after the extension movement had been established on a national basis by the Smith-Lever Extension Bill enacted by the Federal Congress in 1912, by the terms of which federal money is granted to the states for education directly on the farms in proportion to the agricultural population. This federal appropriation must be matched by an equal amount from the State. This extension movement, now grown to vast and intricate national proportions, seems to have had its beginnings in an appropriation by the New York State Legislature to the Department of Agriculture at Cornell University, now the New York State College of Agriculture, in 1895, to carry the teachings and inspiration of the University to the young people and the men and women on farms who could not become resident students in the University courses. This movement had its inception in the mind of John Spencer of Westfield, who later was affiliated with the University, and under the familiar title of "Uncle John" was for years the inspirer of boys and girls concerning the things of the "out-of-doors." The conception of direct teaching by the University beyond the walls of the institution grew from this and other pioneer efforts into the great extension movement.

Other movements worthy of special mention that have especially affected the agricultural interests of the State, and some of which have reached far beyond its bounds, are the Grange or Patrons of Husbandry. The Patrons of Husbandry, commonly known as the Grange, which was fathered by O. H. Kelley at Washington, D. C., had advisors in different parts of the country. The first local chapter, Grange Number 1, was established at Fredonia, April 16, 1868. New

York has always been one of the strongest Grange states in the Union, and has stuck constantly to the movement. It now numbers 862 local granges, 56 Pomona or county granges, and has a membership of 114,000. It has been a potent force for agricultural leadership and inspiration in rural affairs.

The New York State Horticultural Society was formed from the Western New York Horticultural Society and the New York State Fruit Growers Association in 1918 and now has a membership of 1700 which makes it the largest as well as one of the most live associations of horticulturalists on the continent and perhaps the largest in the world. The Western New York Horticultural Society was founded in February, 1855, and for many years looked after the horticultural interests of the State. In 1901, there was a split in its membership, largely along the line of nursery interests and fruit production, and the New York State Fruit Growers Association was the outgrowth of the movement. For many years, annual meetings and one or more field trips have been held by each of these organizations. They included in their membership a very large proportion of the fruit-growers in the State and have been effective leaders in that field. In the State Horticultural Society the membership of the two organizations is now united.

The third movement has to do with the dairy interests and is embodied in the Dairymen's League. Its inception was the desire and a certain degree of necessity for a better price for milk if it was to continue to be produced in quantity at all adequate to the needs of the city population. The League began in 1907. Its financial support consisted of an assessment on each membership of twenty-five cents for each cow owned.

CHAPTER III

SOILS OF NEW YORK

FEW, if any, states have as large a variety of soils as New York. This, when coupled with the existing differences in elevation and in climate and varied nearness to market, forms the basis of a wide range of cropping schemes and of agricultural development.

The soils of almost the entire State have been formed under the influence of glacial processes. As has been stated, the glacial incursion came into the State from the north and covered all but a small angle of country south of the Allegheny River, in Cattaraugus County. The general movement of the ice was deflected by the contour of the land surface. Tongues of ice in the valleys protruded far forward of the main mass. In the Mohawk Valley the general direction of movement of the ice as shown by scratches on the bed rock was nearly east and west, due to the deflection of its movement around the Adirondack Mountains from the Hudson Valley on the east and the St. Lawrence Valley on the west.

In studying the soils of New York, it is important to keep in mind three facts: First, that the ice moved in general from the north; second, that the exposure of the different rock formations lay across the general path of movement of the ice; third, that throughout the southwestern two-thirds of the State, the general slope of the land is to the north. As a result of this condition, ponds and finally great lakes were formed in the hollows between the front of the ice and the northward slope of the land as the ice gradually retreated northward due to melting. Naturally, the water accumulated in the valleys and sooner or later attained a level that permitted it to drain off to the southward in the form of great swollen rivers. The valleys of the Delaware, Susquehanna and Allegheny rivers bear evidences of this former flood level in the remnants of gravel, sand and silt that form terraces along their courses.

As a consequence of the three sets of facts just pointed out, there has been very extensive mixing of the rock material from different formations. In general, the material from each rock formation is most abundant to the southward of its exposure. This feature of the distribution of rock material is especially important to keep in mind in considering the occurrence of limestone which has a very large influence on the crop-producing capacity of the soil. In a glaciated region there is not necessarily any close relation between the underlying rock and the soil material that rests upon it. The soil has probably been brought in from a greater or less distance.

The unconsolidated earth above the bed rock is extremely varied in thickness, in fineness, in content of stone, in color, and content of organic matter, in content of limestone, in drainage, and as a result of

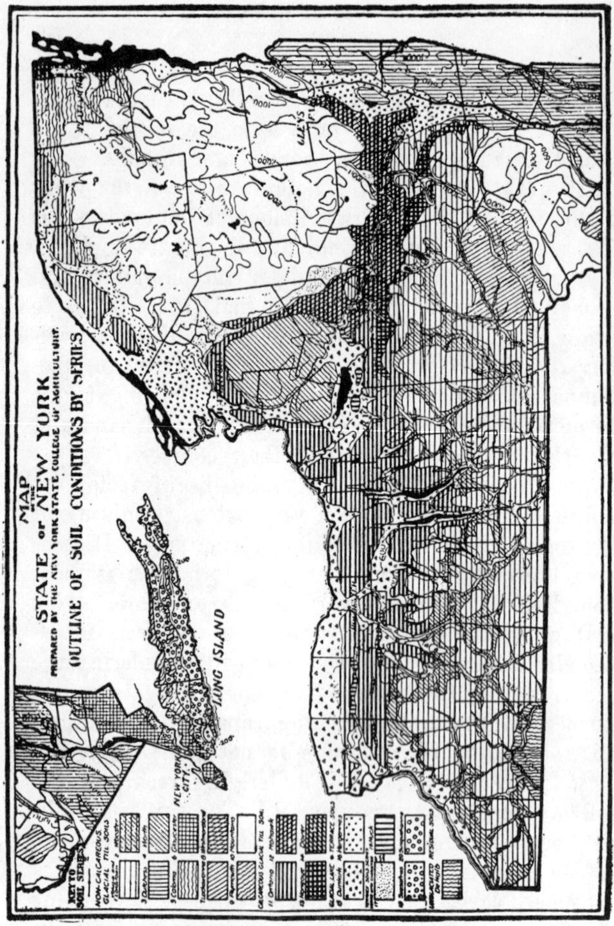

Fig. 17. Showing the distribution of the main series and groups of soil.

all these in crop-producing capacity. In general, the soil formation is from ten to fifty feet thick. It is deepest in hollows and protected coves and thin or even entirely absent over the higher hills and mountains. In the Hudson Valley and in the Adirondack Mountain region, exposures of bed rock are common.

Information in detail covering the soil conditions in New York may be found in the soil survey reports. The survey has covered about 28,000 square miles or about one-half of the State, mostly by counties. The reports on the areas consist of a descriptive pamphlet and a map. The maps represent the occurrence of the different kinds of soils in relation to the geographic features of the country on a scale of one mile equals one inch. The descriptive report treats of the location and general physical features of the area, its geology, climate and agricultural history. Each kind of soil is described in some detail and its relation to crop growth and to the more important soil improvement practices is pointed out. The aim of the soil survey report is to give a comprehensive account of the soils and natural agricultural features of the region.

The characteristics by which the different kinds or types of soil are recognized are much the same as the farmer uses, and include the color of the material, the kinds of rock of which it is composed, the proportion of limestone and organic matter it contains, the arrangement of the material, whether unsorted or sorted and arranged in layers of different fineness, the extent to which it is drained and the general lay of the surface, as well as the fineness of the material and the occurrence of stone.

The individual kinds or types of soil fall into groups in each of which the types have many characteristics in common, such as color, kind of rock material and arrangement. Such a group of soils forms a soil series.

The soil conditions of the State may be described by physiographic regions, as already enumerated. The distribution of the different groups and series is best represented by the map in Fig. 17. This should be studied in connection with the relief map of the State (Plate I).

SOUTHERN PLATEAU

This region of generally high elevation with deeply dissected valleys is mostly covered by light colored stony soils, formed directly by glacial ice. The underlying rocks are predominately gray to black sandstones and shale with occasionally very thin strata of impure limestone. These are embraced in the Catskill group of rocks of the Devonian series. All of the soils of the region are deficient in lime and over large areas this is so pronounced as to be the chief limiting factor in the production of large yields of crops.

All the unglaciated soils are embraced in the DeKalb series. The stony loam type predominates. The DeKalb soils occur exclusively in the southwestern part of the State. In this region the beneficent effect of the glacial invasion is well illustrated by the contrast in soil character and general agricultural development between the glaciated region and the unglaciated part. The contrast is especially impressive a little way southwest of Randolph where excellent glacial soils have been pushed up into close contact with the rough, infertile and undeveloped unglaciated soils of the DeKalb series that lie to the south. The general effect of the ice has been to deepen the soil, smooth the surface outline and to improve both the physical and chemical composition of the soil. The DeKalb soils are light colored, stony, and in texture border on a clay loam. Most of the area is in timber. Further south in Pennsylvania, the DeKalb series occupies a large area and includes types of soil less steep and stony and better suited to agricultural development than those in New York.

From the crest of the Catskill Mountains, westward throughout most of the southern two tiers of counties in New York and reaching westward into Ohio and Pennsylvania, the hill regions are occupied by glacial soils of a uniform character. These soils are generally thinnest, most acid and least fertile on the steep slopes and over the top of the higher hills, and are deepest and most productive in the valleys and protected coves, especially those valleys having an east-west direction. The soils are also relatively thickest and best on the northern boundary of the plateau where it joins the Lakes Plains region.

Five main series of soils are developed in the plateau region, and may be divided into two subgroups — one characterized by yellow-brown and the other by red to pink colors. The soils of yellow to light brown color make up three series. In two of these the soil formation is relatively thin and ranges from two feet to eight or ten feet in depth. These are the Lordstown and Volusia series. The latter is distinguished chiefly by compact hardpan structure in the deep subsoil, resulting in poor drainage. The third series in this sub-group is the Wooster, in which the soils have a larger depth, more material foreign to the bed rock of the locality, and a more complex structural and textural character. The red-pink soils are the Culver series in the upper Catskills, the Lackawanna series in northern Delaware County, and parts adjoining, and the Chenango series of high terrace soils all through the western half of the State.

All of this plateau region originally bore a heavy growth of timber. White pine predominated but there was a heavy admixture of broad-leaved trees that included oak, sugar maple, chestnut, elm and many other species. This section needs much under-drainage in spite of the prevailing heavy slopes of the land.

The fertility of the soil is only one factor to be considered in the agricultural development of this plateau region. The topography of the country and the situation of shipping centers and markets are equally important factors that react on the social and agricultural development. Much popular discussion has been directed particularly at this region of so-called abandoned farms. It has much less population as a whole and especially in the country districts than it had forty or fifty years ago. On the whole, this district has not been able to hold its own in the agricultural competition with the Middle West.

THE GREAT LAKES PLAIN AND THE MOHAWK VALLEY

Stretching northward from the foot of the southern plateau region, across the Great Lakes plain and the slopes of the Mohawk Valley, is a deep accumulation of glacial and glacial lake deposits. Here limestone formations are most extensively exposed, and contributed largely to the ice mill. The great depth of the glacial ice, coupled with its choked movement due to the highland to the south, formed deep masses of till over most of the area and imparted to it surface features favorable for tillage. Later, glacial lakes occupied large areas in the lowlands and the local ice deposits were eroded, reworked and scattered over the lake floor in sheets of clay, silt, sand and gravel. Numerous ponds and lakes were formed that were subsequently filled with mud and by the growth of plants form rich marsh soils.

This is the region of the best soils and the highest general agricultural development in the State. The fertility and favorable physical condition of the soils, combined with the low elevations and beneficent climatic conditions, produce an intensity and diversity of crop production, and a general agricultural prosperity unsurpassed by any other state of the same size, and seldom equalled or excelled on the American continent. The soils belong to two main divisions:

those till soils formed directly by the glacial ice and those subsequently formed by water arising from the melting of the glacial ice during the retreat of its front.

The glacial till soils are the most extensive of the two divisions and occupy all the higher portions of the region. The glacial till sheet underlies much of the glacial lake deposits and often protrudes through the surface. In addition, they reach up the foot of the slopes to meet the soils of the highland areas into which they gradually merge.

For the most part, the glacial till soils are moderately stony. They contain occasional foreign bowlders of igneous rock brought from the north. Bowlders of the local rocks of the region are more common than the foreign ones, and while the removal of these rocks facilitates tillage, that operation is not often necessary in order to utilize the land. The type and proportion of stone vary widely in some sections. Usually in close association with the underlying limestone formations, blocks of massive limestone constitute the predominate field stone. These are most abundant on the southern rim of the Mohawk Valley and of the Ontario plain. In the Mohawk Valley and on the southern flank of the Adirondack Mountains, the hard granite type of rock is more common. Across the northern flank of the Adirondacks and again on the Ontario plain, sandstone is most common. The deeper subsoil of these till formations is usually very compact and may have properties akin to hardpan.

The predominence of different kinds of rock give rise to several series of glacial soil. The most extensive of these is the Ontario which prevails all through the western plains region and eastward beyond Utica. The Medina shales and sandstones, the Salina and part of the Hamilton shales have entered most largely into the formation of this series, together with rather large quantities of limestone from the outcrops of the region.

The lime content is fairly ample, as might be expected from its association with limestone formations. The deeper subsoil is always heavily charged with limestone fragments but in the surface two-to-four-feet this is not so abundant, and that part of the section will seldom show the presence of free lime carbonates. However, it is sufficiently well supplied so that acid sensitive crops such as clover grow very well. On this series the production of alfalfa is most largely developed. The fruit industry of western New York is largely associated with this series of soils, and it is one of the best for mixed farming. Cabbage and beans are largely grown on the same soils. Teasel and hops are special crops developed at Auburn and south of Utica respectively. This same soil covers much of the best live-stock producing sections of the State including the region to the south of Syracuse and extending east and west for many miles.

The widest development of the series is through the upper part of the Finger Lakes region, and thence northward to Lake Ontario. It spreads well southward around the foot and along the shores of the Finger Lakes. Its relation to the heavier movement of the glacial ice is indicated by its furthest southward reach in the Cayuga-Seneca lakes section, which is the center of the Finger Lakes trough. Lower Seneca County has a larger development of this series than any other county of the same latitude, except perhaps Livingston, where the broad valley of the Genesee River similarly favored the southern development of the series. Around the eastern end and on the south shore of Lake Erie, the series is not much developed, partly because of the abrupt slope to the highlands. To the eastward its development is modified by the material from the Adirondacks. Its most northern occurrence is at the east end of Lake Ontario in southeastern Jefferson County, and in Central St. Lawrence County where it is associated with the Trenton limestone.

Standing close to the Ontario series, in character but not in extent, is the Honeoye. This is closely associated with the exposures of the limestone formations in the same district in which the Ontario series occurs. It represents that part of the till region where limestone was most largely introduced into the soil and where there is sufficient lime carbonate, even in the soil, to effervesce freely with acid. The field stone are nearly all limestone. The subsoil is filled with limestone fragments.

The till mantle that forms the Honeoye series is relatively thin and in the stony type limestone ledges protrude in many places. Three types are common in the series, the stony loam, the loam and the fine sandy loam. The two latter are excellent soils. They are sometimes a little shallow, which property interferes with their moisture supply. Their high lime content makes them especially favorable to the growth of alfalfa and clover. Blue-grass thrives, as it also does on the Ontario series. The hop industry, formerly more prominent than at present, and developed from northeastern Madison to northern Schoharie counties, is largely associated with these series.

East and northeast of the Ontario-Honeoye occurrence, through the middle of the Mohawk Valley, the glacial till from the Trenton limestone and from the black calcareous Utica shale is mingled with the igneous material, brought down from the Adirondack region. The Hudson River sandstone and shale on the north side of that valley contribute some material. The result is a rather compact material that ranges from a clay loam, through the silt loam to the loam in texture, and is called the Mohawk series. The limestone and the black Utica shale have contributed a considerable amount of calcareous material. The black shale gives a dark to black color, especially when wet. Finally, the granite and other igneous bowlders from the Adirondacks that make up much of the field stone and in finer form enter into the soil, distinguish this series from the calcareous soils described above.

In crop-producing capacity, the Mohawk series closely resembles the Ontario. Especially is this true of the black loam. It grades off to the adjacent series on the plateau to the south and to the mountain soils on the north. It is preëminently a grain, hay and dairy region, largely because of the character of the climate under which it occurs.

The glacial lake soils exhibit a greater diversity in the character of the types, but a greater uniformity in series characteristics than the glacial till soils. All the glacial formations and all the rock formations of the region were drawn on by the waters of these lakes and by the streams flowing into them for the material that was strewn over the bottom of the lakes. This general uniformity of the soil material is most noticeable in the former beds of larger lakes. Those are adjacent to the present shores of the Great Lakes.

The soils formed in the high-level lakes that occupied the first hollows at the south front of the ice exhibit somewhat more diversity since the material brought in by the streams was more local. However, this is rather a fading out of the typical character of the larger lake deposits than a definite change in character.

The most widely developed lake formed series is the Dunkirk and is made up of light brown or yellow to chocolate brown material that ranges from heavy clay to light sand and gravel. The sand, in small areas, is so clean as to be blown about freely by the wind. As has been indicated, the material represents the assorted wash of all the adjacent higher formations. South of Lake Ontario, the soil has a decided pinkish tinge, probably due to the influence of material from the red part of the Medina and the Salina formations that give rise to the Lockport series. With the dark or black shale, this produces the rich chocolate brown color of the heavier types that characterizes the series.

The boundaries of the glacial lakes were very irregular. The earlier levels to the southward were the higher and reached up to a thousand feet or more. The more gentle slopes and level areas further north are devoted to mixed farming. Fine sandy loam, loam and silty loam types predominate and as is usually the case with these classes of soils, they are about equally valuable for the production of a variety of crops.

The agricultural experiment stations both at Geneva and Ithaca are situated, one partially and the other entirely, on the Dunkirk series, the clay loam predominating. At Geneva the station also includes some of the Ontario fine sandy loam soil.

With the Ontario series, the Dunkirk occupies the bulk of the fruit-producing region. Both are made up chiefly of strong types of soil. Of course, climate, produced by the low elevation and the influence from the Great Lakes, is a factor in the development of fruit. Good farming conditions prevail on most of the Dunkirk series.

Closely associated with the Dunkirk series of soils is the Clyde, which represents that part of the lake deposit that, because of its low position and flat topography, has been in a pronounced swampy condition. Consequently, it has received much rich wash from the adjacent land and has accumulated enough organic matter from the remains of plants to have a very dark or black color to the depth of at least a foot. It differs from muck and peat soil in that it is not predominately organic. Sandy loam, loam and clay loam predominate and make first-class soil for grain, hay, and vegetables. The Clyde soils are most extensively developed on the western part of the Ontario plain in Niagara County.

THE ST. LAWRENCE AND CHAMPLAIN VALLEYS

This region is made up of a relatively narrow rim of tillable land that reaches around the base of the Adirondack Mountains. Its width along Lake Champlain is narrow and ranges from less than two or three miles to ten or twelve miles in the northern part. The tillable land in the lower valley is exceedingly narrow or entirely absent. In the St. Lawrence Valley the width ranges from ten to about thirty miles, and the inner or mountain border is irregular.

The topography and soils naturally subdivide into two divisions: the undulating to flat portion at the lower levels that have been covered by glacial lakes; the higher, rolling to hilly portions covered by glacial till soils. The lake soils are generally quite free from stone and range from heavy clay to light colored sand and gravel. The glacial till soils are moderately to prevailingly stony.

The till soils are subdivided into three main series, based on the predominant character of the rock materials with their attendant physical properties. These are: the Dover series derived from the influence of igneous rocks and sandstone on calcareous till from the Trenton limestone; the Coloma series from the predominant influence of the Potsdam sandstone; and the Worth series from the predominant influence of the Hudson River shales and sandstone in their western development where they are horizontally bedded and but slightly metamorphosed but are coupled with the slight influence of igneous material from the Adirondacks and of some materials from the pink Medina sandstone.

The Dover series is the most important agriculturally of the three derived from till in this region. In general character and crop-producing power, it resembles the Ontario series. It usually has a more decided light brown color and a more open, friable structure.

The soils of the Dover series need more or less drainage. They are also benefited by applications of lime for acid sensitive crops, although the soils are generally strong and durable. The deep subsoil is heavily stocked with limestone fragments and the field stone are largely of that rock. Hay and forage crops are the predominant products. In Franklin County hops were formerly grown mostly on these soils and in northeastern Clinton County apples are produced somewhat extensively. In the production of these two crops, the series shows its relation to the Ontario and Honeoye. In northern Lewis County and the adjacent portions of Jefferson

County, and in fact wherever the series is developed, the possibility for the production of grass, hay and forage crops has led to the extensive use of the land for dairying.

The Coloma series is distributed around the northern flank of the Adirondack Mountain area in association with the Potsdam sandstone. The hard character and resistant arrangement of that massive formation has resulted in a relatively thin stony soil. The tillable areas are small and irregular and occupy protected hollows and slopes in the rock structure. Large and small bowlders are thickly strewn over the surface.

The soil and subsoil are a light rusty brown color. Loam and sandy loam types predominate. Drainage is generally good and the lime content is low. The content of organic matter is also relatively low. Consequently, the general agricultural development is low and without special characteristics. Grass and grain do poorly on the soil. The one crop to which it has been found to be preëminently adapted is potatoes, of which good yields and excellent quality are secured. The physical properties of the soil, coupled with the cool even climate produced by the latitude and the elevation, combine to give it this favorable relation to the potato crop. The region is developing a reputation for seed potatoes for southern planting and bids fair to compete successfully, though on a smaller scale, with the famous Aroostook district in Maine.

The Worth series is recognized on the southwest

flank of the Adirondack Mountain region in lower Lewis, northwestern Oneida and eastern Oswego counties where the Hudson River and Medina shales and sandstone furnish the bulk of the material. In general, the soils are quite stony. The resistant sandstone contributes bowlders that strew the fields and in the cleared areas have been built into fences. The elevation ranges from 1800 feet in the Lewis County portion down to 600 or 700 feet at the lower limit in Oneida and Oswego counties. A large part of the series is in timber and bears a heavy native growth of hemlock, spruce and a few hardwood trees of maple, beach and elm. In the more accessible parts, the timber has been removed and the land remains in birch slashes with a mixture of more useful species. Farm conditions are rather backward and decadent. None of the common crops makes an especially good growth.

A rather quiet lake at some time covered large areas in the St. Lawrence and Champlain valleys. The low rock structure, especially in the former region, was covered by an extensive glacial lake and the sediment laid down over the surface formed a thin covering on the high points and deep deposits in the hollows. In western Jefferson and St. Lawrence counties, innumerable masses of igneous rock and a few limestone ridges protrude through the lake deposits, which condition is well illustrated by the Thousand Island region.

These lake deposits form the Vergennes series and closely associated with them on the mountain side are large areas of the Merrimac series. In the Champlain Valley they reach up to the foot of the steep slopes that rise to the mountains. The deep water deposits which predominate through the region, except near the shore, are silt and clay of a light brown color. They give rise to the clay loam and silt loam types. These form a broad, flat to gently undulating surface, and coupled with the compact character of the subsoil this surface feature produces large areas of wet land in need of under-drainage. The common practice is to use shallow surface ditches for drainage, which are very inefficient. Many of these are formed by plowing the land in narrow beds a rod or two in width with a broad "dead furrow" between.

The utilization of the soils and the general agricultural development of this region is handicapped by the short season and by the limited transportation facilities. The seasons are further shortened, especially on the flat heavy soils, by the poor drainage. Better under-drainage, coupled with the development of special money crops such as potatoes, in addition to dairying, and the adjustment of the systems of farming more effectively to those practices will produce better results from the land.

THE HUDSON VALLEY REGION

Both the soil conditions and the agricultural development of the Hudson Valley region may be defined by the word "piecemeal," and so they must continue in view of the predominance of those "small-scale" practices that go with small operations. The rough topography that generally prevails, together with the dissection of the tillable areas of soil by the numerous rock exposures will continue to interfere with the development, except in limited regions, of large, well arranged progressively managed farm units. Its proximity to large centers of population, together with the beautiful roll and outlook of the country, strongly stimulates the "summer boarder crop." This indeed is spreading in ever increasing numbers up the valley and over the Catskills, and is breaking up enterprises of a strictly agricultural character.

The glacial till soils fall into three series, one of which, the Dutchess, is far the most extensive. It is the result of the glaciation of the Hudson River formation of sandstones and shales. Occupying pockets through the region, the limestone and marble have given rise to the calcareous Dover series, which has a small total development. The igneous and highly metamorphosed gneiss rocks in the heel of the State, gave rise when glaciated to the Gloucester series, the typical hill lands of much of New England.

The three most extensive soil types in the Dutchess series represent differences in general depth and character and amount of stone content. The silt loam is the deepest, the most generally tillable, and by far the best agricultural soil for extensive farming. Hay pasture and timber are its most natural products. When well handled, it gives good yields of grain, corn, wheat, oats, rye and buckwheat. The last three are prominent crops. These crops, with the pastures and corn for silage, are combined with dairy farming which is perhaps the most common industry.

In the middle part of the valley, fruit-growing has attained some prominence and most of the fruit is produced on this type and certainly on this series of soils. Tree- and small-fruits are extensively developed. A further factor in their development appears to be the long growing seasons that prevail here, due to the influence of the drift of air up the Atlantic Coast into the lowland area.

The glacial drainage over the irregular surface of this region formed gravel and sand terraces in all the main channels, remnants of which remain to give rise to soils similar to the Chenango series in the plateau region and of similar crop and agricultural value. There is the same range from loose coarse gravel and sand to sandy loam and silt material.

ADIRONDACK MOUNTAINS

The Adirondack Mountains scarcely require mention as to soils, as the region has small agricultural value outside of forest purposes. This elevated area was traversed by the glacial ice and large areas of the rock raked bare. Their resistant character and the sluggish movement of the ice produced a small amount of material. The drainage water incident to the retreat of the ice further reduced this, and left, in addition to a partial and irregular sandy covering over the slopes, deep sandy fills in the valleys that are crowded with bowlders. The irregular filling of the valleys produced innumerable lakes and through these and down the slopes the streams meander, forming numerous and extensive falls that have large potential water power. In its way the soil covering of the region will produce a large amount of timber, but active farming is justified only under very unusual conditions.

LONG ISLAND

The last physiographic division of the State to be considered, and the lowest in general elevation, is that comprised in Long Island and small areas on Staten Island, together with the adjacent parts of the mainland, considering Manhattan as such. Nearly all of this belongs to the Atlantic coastal plain division of the country, and the soils have been largely formed under the ocean. The great terminal moraine of the glacier cuts eastward from New Jersey and lies across Long Island. But even that has been largely overtopped and cut away by the ocean waves so that most of the area is a gently sloping sandy plain. Heavy glacial till is exposed in scattered areas on the northwest shore of the island in the northern suburbs of the city of Brooklyn. Reaching east from this heavy till is a much more sandy and gravelly till that forms two lines of irregular hills. These cross the island, eastward, one in a diagonal direction to Montauk Point, and the other is situated on the north shore of the island and known as the Harbor moraine. They rise to a maximum elevation of 400 to 600 feet at different points. The rough surface of these interfere with their utilization for agricultural purposes. The north shore is generally quite precipitous and is indented by many shallow bays. The sand plains that in many places cut through the glacial formations spread out to the south to seashore beyond which are mud flats guarded by an irregular outlying line of sand bars. Much of the material of this plain was derived from the adjacent glacial formations which in turn were brought down from New England by the ice.

Thousands of acres of land all through the middle of the island are in a state of near-wilderness. It is covered by scrubby timber, is unsettled and unfenced except as limited areas have been developed for suburban residences, and the roads are poor except a few main highways. It is open to the wild birds, the rabbit and the deer, in spite of its proximity to the greatest center of population on the continent.

The western end of the island within twenty-five miles of New York City has a good loam to sandy or silty loam. It has the depth and body that make most of it suited to the large development of the market-garden industry of that region.

The marine plain has been largely formed from the destruction and distribution of the glacial moraine. A peculiar development is the Hempstead plains soils in the town of Hempstead, reaching in several irregular areas from west of Mineola to near Farmingdale. The soil has a dark to black color and a moderately productive quality. The central portion of this area, which is a slightly sandy to silty loam two to five feet deep over the gravel, is one of the few examples of a native prairie area east of the Appalachians. This is known as the Hempstead plains. Its original vegetation was a rank growth of sedge grass and was early used as the common pasture. Many kinds of vegetable crops grow well on this soil. Potatoes give particularly good yields. The transition to the adjacent poorer sand plains is through a gravelly loam of lighter color and lower crop value than the Hempstead loam. The cause for the dark color and treeless condition of the area has not been explained.

On the eastern end of the island, the soil and consequently the agricultural conditions are again much better than in the middle of the island. The soil has a finer texture and a better body. The sandy glacial material that forms the core of these two eastern prongs of the island is mixed with considerable silt and clay and reworked into a fairly smooth surface so that it makes good farm land. This soil has proved suited to the production of early potatoes. Cauliflower, Brussels sprouts and Lima beans are also prominent crops, their production centering roughly in Riverhead. Considerable corn is produced and on the southern prong of the island dairying is still practiced.

The island is dominated by its availability for residence purposes and its value for home-making has often misled individuals as to its agricultural possibilities. For legumes, practically all the soils need lime. Lime is beneficial to many other crops. On these soils the chief objection to its larger use is the possibility of favoring the disease known as scab on potatoes. Shore swamps and the limited areas of heavy soil need drainage. Humus is generally deficient in the soils. Because of the physical nature of the soil, and its poor relation to moisture for crop growth, and also because of its proximity to so large a population, irrigating with sewage should be considered wherever conditions are favorable.

SPECIAL SOILS

The stream bottoms and the marsh and swamp lands of the State constitute a division of land that is essentially distinct because the character of the soils is much alike through the different areas.

Along nearly every stream, small or large, is the ribbon of first bottom alluvial soil, traversed by the stream either in a meandering sluggish channel or with a direct rush. The width and quality of the bottom land is generally proportional to the former condition — sluggishness. This means frequent overflow as a result of which the sediment is deposited. The nature of these bottom lands is much affected by the character of the adjacent lands from which the wash is derived. However, this quality is considerably masked by the mixing of the material, together with the large and deep incorporation of organic matter. Usually, the presence of organic matter is evident to the depth of three feet or more.

Plate III. Lettuce growing on muck soil. An example of intensive vegetable culture common on the numerous areas of muck soil.

A dark gray to brown color prevails. The larger part of the first bottom soils is placed in four main series that are distinguished chiefly by their color. The lightest colored of these is the Caneadea series. That of medium dark color is the Genesee series which is the most extensive. The red series is called the Barber and is associated with the Lackawanna soils. The dark or black alluvial soils form the Papakating series.

The prevailing class in all these series is the silt loam. Along small streams the loam, sandy loam and gravelly loam are more common.

The surface is likely to be mildly cut up by old abandoned channels and overflow courses as most of the land is subject to annual overflow, by which additions of sediment are formed. The bank of the stream is usually highest and coarsest in texture. The foot of the slope to the upland is lowest and most poorly drained. These soils make first-class corn, hay and vegetable lands. Small grains are liable to lodge and to injury by overflow. Summer crops are safer than winter ones.

Swampy soils have been formed in many blind hollows and pockets in the surface of the country.

The glacial incursion has given rise not only to hundreds of lakes of all sizes and shapes, but to an even larger number and variety of swampy areas and mud flats. The low south shore of Long Island and a little of Staten Island represent the marine type of tidal flat which must be reclaimed if at all by diking, and perhaps pumping the drainage water.

This has been done only to a very slight extent. Such land is saline and usually bears a dense growth of grass. However, a few seasons of protection from the tide, coupled with leaching by the rains would make the tillable land suitable for crop production.

The inland fresh water swamps, native and reclaimed, have an aggregate area of about 2000 square miles. A number of these, including one of the largest, are situated in the Hudson Valley. There are a number of important areas in the Champlain and St. Lawrence valleys and in the southern plateau region, but the greater part in number and area are in the Great Lakes plain. From Oneida and Oswego counties west to Buffalo and south into the Finger Lakes region, large and small areas of swamp and marsh land dot the surface. The intervals between the drumlins are a common position. The small areas are usually deep and are kept quite wet by springs on the border. The large areas are shallow sheets where border springs are not so well able to maintain the water level.

The larger areas of muck soil are in the Wallkill marshes in southern Orange County; the Cicero and Rome areas near Oneida Lake; the Montezuma marshes on Seneca River; the Oak Orchard area on the north line of Genesee County and the Pope Mills area east of Black Lake in St. Lawrence County. A large number of small areas lie among the drumlins in Wayne ounty and in the hollows in the surface of Oswego County.

A large part of this land is still undeveloped, except for pasture. Much of it is in timber. Favorably situated areas are under cultivation. The black swamp soil, Clyde, and the muck are excellent for the growth of vegetables, as celery, lettuce, onions, spinach and cabbage, for forage crops, corn, and for hay, timothy. Large yields are derived. These lands constitute one of the important potential resources of the State, but since they are special and not general crop soils, and their development is an expensive process, it should be undertaken with caution. Special fertilization is necessary, particularly on the muck, and special knowledge and training go with success on such soil.

CHAPTER IV

OTHER RESOURCES OF NEW YORK STATE

TREES grow naturally throughout New York as they do also in New England. They spring up freely in every neglected field and fence corner, as grass does on the western prairie or moss in the cool forest shade. The farmer contends with brush — the young trees — as he does with common weeds. Given a chance they will make every fence-row a dense hedge of vegetation, overgrown with vines. To the person accustomed to the difficulty of growing trees in the middle and western states, this ease with which trees spring up in New York is a matter of wonder. Their cool shade reaches into the villages and many of the larger cities and gives them a cool and comfortable aspect.

Practically every part of the State was originally covered by forest, much of it very dense. About the only exception was the prairie area of Hempstead soils on western Long Island. Great forests of pine and hemlock mixed more or less with hardwood species covered all the lower and better lands and reached well up into the mountain regions. In the latter, spruce and hemlock were more common. The quality and variety of timber reflected the quality of the soil and the nature of the climate, as is the case with the cultivated crops that succeeded the forest. Indeed the original settlers were guided more or less in the choice of their land by the timber growth it bore.

But to the pioneer farmer bent on producing things to eat and wear and sell, wheat and flax and wool, the trees were mostly an encumbrance to be got rid of — branch and stump and root — as rapidly as possible. The market for timber was small, especially as compared with the supply. Hence, one of the earliest sources of cash to the pioneer was the sale of "black ash salts," potash salts, presumably for the arts and for the manufacture of gun powder. The timber was burned on the ground and the potash salt extracted by leaching the ashes.

A little later New York became the leading state in the production of timber. In 1860 Albany was the most important timber market in the country. In 1914 New York State stood twenty-third and is going down, although still leading in farm forest products.

The forest resources of an old settled state have a direct relation to the soil conditions. The hand of the farmer crowds back their border from the better to the poorer or less accessible land. Forest land is thin, steep, rough, poor, or swampy, and does not lend itself to the plow. In New York the zealous pioneers who came largely from an even rougher country, New England, reached over the border of the arable land, encouraged by a different set of prices and market conditions than now prevail, and actually brought much of this natural forest land under the plow. Fortunately, their successors have, to a considerable extent, perceived their mistake and are letting such land go back to a timber covering.

The forest and timbered area of the State divides into two parts: (a) The general forest largely owned by the State by original title or reversion or purchase. This covers the greater part of the two mountain areas, the Adirondacks and the Catskills. These are the State forest reserves and are bounded by the fire patrol line. Within these areas the State by a system of look-outs and systems for fire fighting aims to protect the timber from fire. (b) The farm woodlot areas made up of the small or large patches of timber on the lands included in farms and associated with tillage. For the most part this represents the rougher, poorer or more remote lots on the farm and is a compromise with the native instincts of every man to have a bit of forest to draw on for timber and fire-wood.

The mountain forest areas comprise 7,500,000 acres, of which the State owns about 5,000,000, and private interests own 2,500,000. In the Adirondack region comprising 1,800,000 acres, the ownership is distributed as folllows: State 48 per cent; lumber and pulp companies 23 per cent; private parks 15 per cent; improved land 6 per cent; private forests 6 per cent; mineral land 2 per cent.

The farm woodlots have an area of 4,500,000

acres. There are 300,000 acres of virgin timber, about 400,000 of bare land, probably rock outcrops, and about 5,000,000 acres of cut and burned over land. The remaining area of 6,300,000 acres is in various grades of timber in State and private forests.

The average stand is estimated at 4000 board feet to the acre, making a total of 25,000,000,000 board feet, to which is added the equivalent of 30,000,000,000 board feet in the form of round timber and fire-wood. The forests of the State now produce about 25 board feet to the acre a year, or a total production of 300,000,000 board feet. The annual cut of lumber in 1912 was 1,000,000,000 board feet, a reduction of about one-third in the five-year period from 1908. In 1917 it had decreased to 360,541,000 board feet, and 913,169 cords (128 cubic feet) of round wood, making a total equivalent to 861,870,781 board feet.

The annual report of the State Superintendent of Forests under the Conservation Commission records the annual cut of timber for 1912 to be distributed as shown in Table II which gives a good idea of the species of timber that prevail, arranged in the order of volume:

TABLE II.— DISTRIBUTION OF TIMBER CUT BY SPECIES

1. Hemlock	**128,440,828**	**4. Spruce**	**52,061,700**
2. Pine	**78,221,480**	**5. Beech**	**41,478,550**
3. Maple	**78,103,985**	**6. Basswood**	**29,703,865**

7. Birch	31,906,350	16. Gum	179,630
8. Oak	25,799,650	17. Cucumber ..	124,800
9. Chestnut ...	18,139,275	18. Butternut ..	121,785
10. Elm	14,766,535	19. Cedar	77,950
11. Ash	11,130,065	20. Willow	57,984
12. Cherry	3,585,555	21. Locust	30,700
13. Poplar	1,567,910	22. Tamarack ..	20,000
14. Hickory	1,386,180	23. Black Walnut	9,525
15. Balsam	237,100	24. Sycamore ..	2,050

Total lumber ....517,205,872 feet board measure
Pulp wood in cords equal to 279,265,320 feet board measure
Round wood in cords equal to ....146,174,077 feet board measure

Grand total ....942,645,269 feet board measure

As a result of the various soil, climate and timber successions, twenty-five types of forest are recognized in the State by the professional forester. These turn on the species of trees that predominate and those that are present as accessories in the mixture. Such a variety of timber species and types afford opportunity for a broad and excellent study of plant ecology. The relation of crop to soil is of course very direct, but the crop, whether timber or tilled, represents the results of so many forces and conditions that the individual is often misled in his conclusion as to the controlling factors in the environment.

From the figures, it is evident that the consumption of timber in the State, whether gauged by the present rate of cutting or by the actual consumption, far exceeds the present rate of production. When it is remembered that about 40 per cent of the area of the State is probably better suited to the production of timber trees than any other crop, it is evident that the forestry problem in New York is one of large importance. The annual production of about 25 cubic feet to the acre is relatively low as compared with the results achieved under modern methods of forestry management in European countries. In Germany, for example, as much as 100 cubic feet is produced in some of the best forests and the average production to the acre has increased under good management from 25 cubic feet to 67 cubic feet. A cubic foot is equivalent to about 12 board feet.

As compared with most other states, New York has taken an advanced stand on the preservation and extension of the forest area. Much as has been accomplished in the last ten years, much more remains to be done to make the forest area produce up to its limit and to bring its management up to the high standard that is possible. The prevailing attitude toward forests and timber is not surprising when it is remembered that within the lifetime of persons now living such a large part of the country was covered by timber that the supply seemed almost inexhaustible. It was not a crop to be planted and matured but an enemy to be contended with. Like all other phases of agriculture, the destruction of the forest crop is now followed by the pinch in the supply and it is not unreasonable to expect that in a few years every commonwealth and its subdivisions, as well as individual farmers, and especially companies interested in forest products, will look on timber as a regular domesticated crop to be planted, cared for and harvested in approved fashion when it is mature. At this present stage the country is passing out of the usual period of extravagant waste of a vast natural resource into one of intelligent economy.

The State is gradually promulgating three or perhaps four lines of policy for the benefit of forest development. The first of these is the ownership and reforestation of cheap land; second, the protection of forest areas, especially the State forest areas from damage by fire; third, encouragement and aid to individuals to extend and develop their forest areas; and, fourth, the provision of educational facilities relative to forest management and forest products.

As has been indicated, the State owns about 5,000,000 acres of land mostly in the two larger mountain parks, the Adirondacks and the Catskills. It is continually acquiring title by tax default or otherwise to small areas of land over the State but mostly in the main forest regions. Through the State Conservation Commission it is the aim to manage this forest area in the most approved manner and there should be derived therefrom a regular crop of timber. This policy is not in working order and is prevented from operating by the provision in the constitution of the State against the sale of timber from State land. This "save-the-talent-in-a-napkin" policy has been discarded in the federal policy in managing national forests. Progressive and efficient management insures substantial returns to the government at the same time that the forests are maintained as a regular crop and it puts into effect a sane policy that exemplifies true conservation instead of putting a premium on inexcusable waste.

In order to protect large forest areas against fire, the State has marked off the boundary of two large areas, one in the Adirondacks and the other in the Catskills, within which special provisions are made for detecting and extinguishing fires. The townships within this area are called State fire protection towns. The areas are subdivided into districts and forest rangers with fire wardens and observation stations equipped with an observer, telescopes and other instruments, and with which telephone connections are maintained. Of course fires are more likely to occur at some seasons than at others. Records show that most fires start in May and August. In the bad season, from sixty-three to seventy fire rangers are maintained. In the best season this number is reduced to twenty-five or thirty. There are forty mountain observation stations. These have had a material effect in reducing damage by fires. State regulations backed by laws have been enacted requiring precautions on the part of individuals who may be in forest areas against letting fires get a free range. Railroad locomotives must have equipment against setting fires by means of sparks or ashes and a system of inspection is maintained. A top-lopping law is designed to hasten the removal of brush by requiring that it be lopped down in close contact with the ground so that it will decay rapidly. It must also be piled so as to hinder the spread of fire.

Campers, sportsmen and others are required to observe certain precautions in making fires and are subject to penalty of fine or imprisonment, or both, for carelessness with cigars or others dangerous materials. Camp fires must have a cleared space ten feet in radius beyond their limit and must be completely extinguished before being left. All these precautions and provisions have resulted in reducing the number of fires started. Most of the provisions for detecting and stopping forest fires are of recent promulgation. In 1918, there were 398 fires reported. Of these 100 were started by smokers, 47 by fishermen, 111 by locomotives and 24 by campers. Ninety per cent are classed as preventable. The efficiency of the fire-fighting arrangements is indicated by the figures showing the number of acres traversed by fires as follows:

1903	464,189 acres
1908	368,072 acres
1913	54,792 acres
1918	7,354 acres

Assuming that this land had the average stand and that all the timber on the burned area was lost, the timber lost by fire has decreased in the ten-year period equal to one and a half times the annual cut in the last named year, and assuming a value on the stump of five dollars a thousand, the saving would be $5,000,000. The cost is less than $100,000.

The exemption or reduction in tax on timber lands

falls into two classes: small woodlot areas of less than fifty or one hundred acres where both land and timber are exempt from tax for thirty-five years, which land must be a stated distance from cities or villages; poor lands assessed at five dollars or less an acre and of any area above five acres. The assessed value of the land not to exceed ten dollars may not be increased for the period of thirty-five years. The stand of trees must be brought up to 899 to the acre and must be subject to the inspection and direction for maintenance of the proper officers of the State Conservation Commission. By these provisions, the owner is protected against burdensome taxes until the time when the crop has come to a fair maturity. The State maintains a number of nurseries from which trees are supplied for planting the State land, and are also sold to individuals at cost. These plantings are now beginning to assume some considerable proportions and will undoubtedly increase more rapidly as the public becomes acquainted with the tax exemptions, the aid furnished by the State and the value of the first crop.

The State maintains three institutions that give instruction in the principles of modern forestry. The division of forests of the State Conservation Commission publishes bulletins on modern forest management and gives public instruction through popular lectures. The more systematic instruction of university grade is given in two State colleges. These are the Department of Forestry of the State

College of Agriculture at Ithaca, and the State School of Forestry at Syracuse, both of which have forest tracts for practical instruction.

FISH AND GAME

Nearly all the species of game animals and birds in the State are reported to be on the increase thanks to fairly comprehensive laws dealing with open shooting seasons and restrictions as to the number of individuals that may be taken by any one person. From time to time certain species may be protected from molestation for a period of years to permit their increase. Of course, game laws are effective in the present stage of society only in so far as they are backed up by a system of wardens and inspectors for their enforcement, since public sentiment is not sufficiently well developed for that purpose. The State is divided into thirteen districts with a chief game warden and a half dozen or more assistants and the restrictions limiting fishing in particular are fairly effective. Some varieties of game — the deer for example — have increased in such numbers in the more heavily forested sections as to be something of a nuisance to farmers because of their injury to crops.

The diversity in timber and stream-cover and in climatic conditions favors a large variety of animals, birds and fishes. The bird life of the region is especially rich, particularly the migrating species that to a considerable extent follow the cover of the Allegheny Mountains in their annual movements from north to south. The timber and swamp covering in the bottoms and on the sides of the main north-south valleys form natural roadways of migration. The numerous lakes afford a congenial home for water and shore birds and within recent years certain species of duck that normally nest and rear their young far to the north, are now found to spend the entire year and to propagate on the waters of the State.

The extensive arrangements for the preservation of bird life and the dissemination of interest in and study of outdoor life, by means of Audubon societies and especially as a result of the introduction of nature-study in the schools, have resulted in much increase in the number of birds due to their better protection. The Weeks-McLain law which became effective March 4, 1913, gives federal supervision over the taking or killing of all birds of migratory habits and has supplemented many of the State regulations relative to them, at the same time making the provisions more effective through the local support the measure receives.

The State has instituted four game farms, the oldest being at Sherburne. These are used for the propagation and dissemination of important species of game birds and mammals. The pheasant has been most extensively distributed in this way. Other farms are at Brownsville in the Adirondacks, and on eastern Long Island. In connection with the New York State College of Agriculture at Ithaca a game farm was established in 1916 with particular reference to the investigation of problems connected with the propagation and care of such animals.

The utilization of the waters of the State for the production of food and game fish as a definite phase of agriculture is just beginning to receive attention. J. G. Needham, who occupies the chair of Limnology and Biology in the State College of Agriculture at Ithaca, points out that there is quite as much, if not more possibility, for the production of animal food in water forms as in land forms and that this type of agriculture is now essentially neglected. The numerous and fairly continuous streams and the many lakes and swampy areas in the State afford excellent facilities for this type of industry. In the institution just mentioned the study of the habits and methods of propagation of these forms of animals and of the organisms and materials used for food by them, and instruction in such matters, is now a regular part of the curriculum.

The state maintains twelve fish hatcheries distributed as follows:

1. Adirondack, Saranac, Hamilton County.
2. Bath, Steuben County.
3. Caledonia, Livingston County.
4. Cold Spring Harbor, Suffolk County.
5. Delaware County.
6. Linlithgo, Columbia County.
7. Warrensburg, Warren County.
8. Chautauqua, Chautauqua County.
9. Dunkirk, Chautauqua County.

10. Fulton Chain, Herkimer County.
11. Oneida, Oneida County.
12. Ogdensburg, St. Lawrence County.

These stations hatch and distribute a large number of species of fish and other water animals suitable for food. In 1913, the total number of fish distributed was 1,287,255,120. In 1918 the number was reduced to 396,319,251. Most of these were put into the streams and lakes. New York is said to lead other states two to one. In 1913 thirty-nine varieties of fish were distributed. The leading varieties were several forms of trout, black bass, river herring, lake and tullibee fry and pike perch.

The state owns and leases 31,665 acres of shell fisheries, mostly oysters. These are located on Long Island Sound, Raritan Bay, on the south shore of Long Island and in the mouth of the Hudson River.

MINES AND QUARRIES

New York is not a leader in the product of mines and quarries, and yet it occupies a place of considerable importance. A large variety of materials is produced and in some of these the State is preëminent. In point of value, the most important ore product is iron. The last available figures are for the census year of 1909. At that time the iron ores were valued at $3,095,023, nearly all of which came from the magnetite ores of the Adirondack region. New York is the only state besides Pennsylvania that produces all four kinds of iron ore, magnetite, hematite, limonite and carbonate. The last three, however, are of small commercial importance. The Clinton ores from the Clinton formation are the most important of this latter group and are related in geological age to the famous ore deposits at Birmingham, Alabama. No other metal assumes importance, although specimens are found.

Next to iron ore in value stands limestone, derived from several formations in different parts of the State. Their total value is a little more than $2,500,000. Blue stones for flagging and building, sandstones for building and especially for abrasives are produced in the southern and the western Catskill regions respectively. Marble of several grades and colors is quarried at Gouverneur and in the central Hudson Valley. Slate for roofing occurs in Washington County where it has been formed under the influence of metamorphism of shale in the same way as marble in that and other regions was formed from limestone as a result of great pressure and heat. Granite quarries are operated in the eastern Adirondacks and trap rock quarries in the Palisades district. Talc and soapstone of high quality are secured at Gouverneur. The Adirondack region also produces feldspar, a little vein quartz, garnet and graphite. The production of graphite by the mines in the Ticonderoga district makes New York the leader in that product. The garnets from Essex and Warren counties are used as an abrasive. Peekskill, in Putnam County, is an important center for the production of emery.

Standing next to limestone in value is salt, which may be regarded as a manufactured product. The

Salina formation has a series of heavy salt beds in its lower part that are drawn on mostly by wells in the counties from Onondaga through the Finger Lakes region to the Genesee Valley. In the latter region at Retzoff, rock salt is mined by means of shafts about 1200 feet deep. New York has long been a leader in this product, recently outstripping Michigan which formerly stood at the top of the list.

Close to salt in value is petroleum. It comes from wells that reach the Devonian sandstones in the Olean district. Gas is found in a series of pocket-like areas in the western third of the State. The gas-producing areas have been estimated to have a total area of 550 square miles. Gas comes from a variety of formations. As early as 1821 a well was drilled at Fredonia and the product used for lighting. In 1841, natural gas was used in the manufacture of salt for evaporating the liquid. These items are of interest since they antedate the general use of this fuel in the states to the west and south that have so far outstripped New York in later years. Another important product in which New York has been prominent, and in which the manufacturing operations have a large place, is cement. In the production of the water-lime or natural cement rock that preceded the use of Portland cement, New York was preeminent and has four centers, Rosendale in Ulster County, Akron in Erie County, Fayetteville and Manlius in Onondaga County, and Howes Cave in Schoharie County.

The Portland cement which is usually a more

satisfactory product because of the control that may be exercised over the material used, has now largely supplanted the natural cement rock and in some places the same formation is used in the new process. Limestone and either shales or pleistocene clays of several ages are used and New York still has a prominent place in the cement industry.

Gypsum is available from the top of the Salina formation over a band of country extending from Utica to the Niagara River and is extensively mined west of Rochester, where it was formerly much used as a fertilizer.

Other products are molding and glass sand and clay for the manufacture of brick and tile. The molding sand and the glass sand are pleistocene — recent deposits. The former is most abundant west of Albany and in southeastern Saratoga County. Glass sand which occurs as a surface deposit north of Oneida Lake is now unused. In the main, the clays are superficial deposits and are suited only to the manufacture of the coarser grades of building material. There are some shale deposits suitable for this purpose, for example, at Alfred in Allegheny County where the State School of Ceramics is located. In these and particularly the manufacture of brick and tile the State takes a high rank, namely fifth in total value. The lower Hudson Valley is by far the most prominent district. None of the higher grades of fire or pottery clay or shales is available.

The total value of these several products in recent years is as follows:

TABLE III. VALUE OF MINING PRODUCTS IN NEW YORK

Clay products	$11,871,949
Iron	3,095,023
Salt	2,897,000
Petroleum and natural gas	2,688,996
Cement	2,409,000
Limestone	2,656,142
Bluestone	910,054
Slate	99,827
Traprock	755,128
Granite	444,435
Marble	344,981
Feldspar	47,166
Gypsum	1,048,403
Talc and soapstone	314,724
Garnet	101,920
Emery	16,389
Graphite	138,905
Other products	184,294

WATER SUPPLY

The numerous streams, springs and lakes in the State, coupled with the fairly generous rainfall, afford a good general supply of potable water. The runoff of about 50 per cent of the rainfall ranges from less than twenty inches in the Great Lakes region to a little more than twenty over the plateau and the Hudson Valley sections, and thirty inches or more in the Adirondacks. Limited investigation indicates that with the extension of the tilled area and the reduction in the area of forest land and forest covering, there has been an appreciable drop in the general level of the ground water-table. As represented by the level of water in wells at different periods, the extent of the drop varies in different regions of the State. That the consequence of this drop is as serious to agriculture as is sometimes believed is questionable, since the moisture used by crops is held in the upper soil by capillarity and is not drawn directly from the body of ground water.

The range in elevation of the land which reaches up to four and five thousand feet in the mountain regions affords a large quantity of water power, as this drainage water passes over the succession of falls and rapids to sea level. Nearly all the larger streams develop some potential water power on their course. The flow fluctuates widely at diffierent seasons in the year. A comprehensive study has been made of the possibilities of storing the flood flow to develop water supplies and hydro-electric power. It has been determined that there are thirty-nine suitable reservoir sites distributed as follows: thirteen on the Upper Hudson; three on the Sacandaga; nine on the west branch of the Hudson; eleven on the Raquette, and three on the Genesee River. One of the facts that is often overlooked is that water supply is one of the important limiting factors in the multiplication of population. It comes even before food and is more critical. Several of the reservoir sites have an immense capacity. The reservoir on Aesopus Creek in the Catskills, which is being developed by New York City, has a maximum capacity of 124,000,000,000 gallons, which would make a body of water fifty feet deep over an area of 5000 acres. That on the Sacandaga at Conklingville has a potential capacity of 32,000,000,000 cubic feet, equivalent to 1150 acre feet. By a dam on the Genesee River at Portageville, 18,000,000,000 cubic feet, equivalent to nearly 700 acre feet may be stored. In one mile a pressure head of 200 feet is available. Many of the reservoir sites are in the State forest parks, and it is now provided that 3 per cent of the area of such state land may be utilized for reservoir purposes.

The hydro-electric power that may be developed in the State is estimated at a million and a half horse power. Of this amount, a little over a half million horse power is in use within the State, to which should be added two hundred thousand horse power developed on the international stream, the Niagara River. It is interesting to note that the total water horse power now in use in the State is about the same as it was in 1824.

The development of water power comes under the supervision of the Conservation Commission. Very little has been done to make this power available to the public and thereby conserve the supply of coal and oil. That it ought to yield a considerable revenue to the public treasurer is generally accepted. At the same time the investment in existing power plants should be reasonably protected perhaps by State acquisition. The most extensive region for power development is around the base of the Adirondack Mountains on the several streams radiating out of that center of high elevation and large rainfall.

CHAPTER V

AGRICULTURAL INDUSTRIES, PLANTS AND CROPS OF NEW YORK

THE crops produced by New York are exceedingly varied. The variety in soils, in climate and in market conditions encourage diversity in crop as well as animal products. The State is exceeded in this variety of crops only by California, which is much larger in area, in climatic reach and in range of elevation and topography. The range in these three primary factors produces an exceedingly complex pattern to which farm practice is continually adjusting itself and the process will doubtless continue for many years to come.

In value of all farm crops, the State ranks eighth. New York produces 3.8 per cent of the total value of all crops grown in the United States, while her proportion of the total acreage in crops is 2.7. From this it appears that the intensity of her production is above the average. This amounts to an average value of $25 an acre of crops to be compared with slightly less than $18 for the entire country as valued in 1909. The rank in acre value of crops is thirteenth. In value of crops to a farm New York is sixteenth with the sum of $950, the average for the country being $860, or nearly $100 less. (See Figs. 18, 19.)

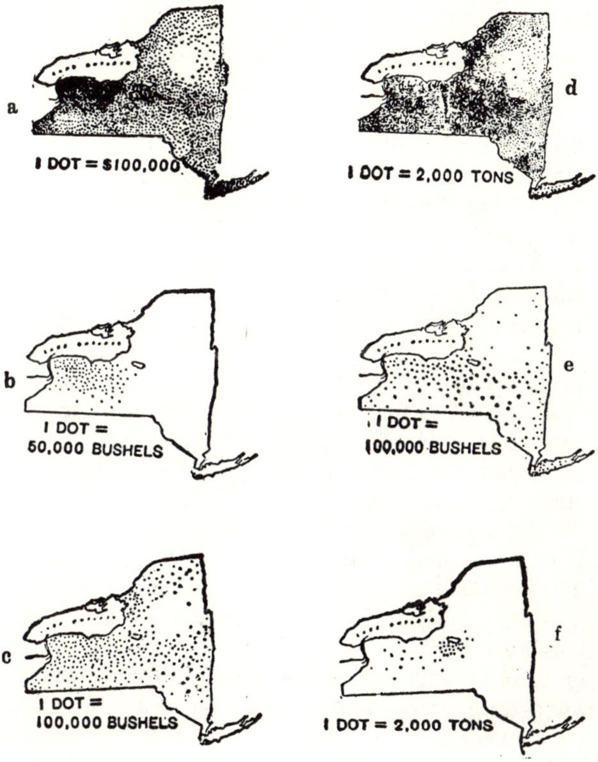

Fig. 18. Total value and distribution of all crops and the amount and distribution of the specified crops in 1909. a. value of all crops; b. wheat; c. oats; d. hay and forage; e. corn; f. alfalfa.

New York ranks first, in the thirteenth census returns, in the production of hay and forage, potatoes, vegetables, buckwheat, flowers and plants, nursery products, small-fruits, willows and teasel; second in orchard fruits and grapes, hops, maple sugar and sirup and ginseng, and until very recently it was third in dry beans. In the production of the grains and cereals it ranks low, having long since given way to the central states. In the last census period, the chief change in its position has been a further depression in rank in the production of cereals, and complete loss of standing in the growing of sugar-beets, while in the production of buckwheat it advanced from second to first place. Acreage is a better measure of the relative production of crops than is yield, particularly in making comparisons between different regions and periods. Nearly 8,500,000 acres or 28 per cent of the entire area and 56.5 per cent of the improved area are devoted to cultivated crops. Three-fifths of this crop area or about 5,000,000 acres is occupied by hay and forage crops, a little less than one-third, 2,800,000 acres, by grains, cereals and seed crops, one-twentieth or about 400,000 acres by potatoes and vegetables, and one-fourteenth or about 600,000 acres by fruit and nursery crops.

The total value of all crops including fruits and flowers in 1909 was $209,000,000. In rank according to value New York stood somewhat different from the acreage relations. The value of farm crops alone in 1909 was approximately $132,000,000. In 1918 it amounted to $281,000,000, though the acreage was then very little larger than in 1909. Fruit moved up from fourth to third place between 1899 and 1909, changing with vegetables.

In addition to the major crops, the production of several minor crops has developed in localized areas. The most intensively cropped portions of the State are, first, a broad region lying south of the Great Lakes, second a narrower section along the axis of the Mohawk River and third the lower point of the State in the Hudson Valley from Albany southward to Newburg. The remaining regions of important agricultural development are fourth, the southern tier of counties; fifth, a fringe of country ten to twenty miles wide around the eastern end of Lake Ontario and along the St. Lawrence River and the Canadian line; sixth, together with a still narrower fringe along the west side of Lake Champlain and through the Hudson Valley above Albany.

The region south of Lake Ontario centering around Rochester and reaching to the Niagara River is the most highly developed and the most thoroughly tilled. It loops southward into the Genesee Valley and follows the general sag in elevation along Seneca and Cayuga lakes. This as a whole is the "flower of the State" in agricultural development. Other regions are as intensively developed over a small area, for example, the grape belt south of Lake Erie, portions of the Mohawk Valley, some of the territory between Albany and Newburg in the Hudson Valley and the east and west ends of Long Island. But when both area and intensity are taken into account, few districts anywhere in the entire country surpass this south of Lake Ontario for a width of forty miles in general intensity of agricultural development.

Hay and forage crops are widely distributed in the tilled area. Pushed aside to a considerable extent by the development of the other groups of crops, they have still maintained the major development, and in conjunction with the market facilities, support the large live-stock and dairy interests of the State. The cereals, except buckwheat, are most concentrated in the Genesee Valley region. Buckwheat is largely a poor-land crop and has the largest area but not the largest yields in the rougher and more remote regions. Vegetables, including potatoes, have several centers of production that in part coincide with particular crops. Potatoes, however, are most largely grown in three centers: west central New York from Lake Ontario southward into the northern half of the southern tier of counties; on eastern Long Island, and in northern New York around the north flank of the Adirondack Mountains in Franklin and Clinton counties. The more detailed distribution and relations of individual crops will appear as the discussion proceeds.

In Table IV is a summary of the changes in crop area in acres in New York State since 1844.

HAY AND FORAGE (See Figs. 18, 20)

In the production of hay and forage, New York should be compared with southern Wisconsin, southern Minnesota and eastern Iowa. Two factors unite

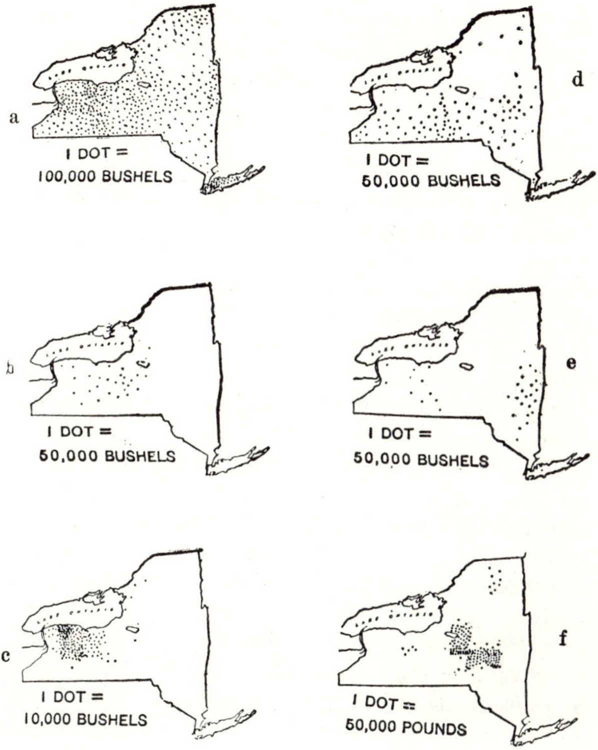

Fig. 19. Amount and distribution of the specified crops in 1909. a. potatoes; b. barley; c. dry peas and beans; d. buckwheat; e. rye; f. hops.

to cause the large yield of these crops. As a cash crop, they are bulky and therefore expensive to ship long distances. New York City and other large cities adjacent to this territory require large quantities of forage and the sale of the crop has been one of the most profitable enterprises on the mixed-crops farm. This fact is shown by farm management figures. Consequently, many farmers make a business of selling hay and if it were not for the limiting factor of soil fertility as commonly handled, many more would follow the same practice. Then, too, there are large areas where the soil, climate and labor conditions make hay the most satisfactory crop, even if only small yields are secured.

Timothy and clover constitute nearly three-fourths of the hay area, timothy occupying the second largest area, somewhat less than one-fourth of the total, or 1,078,358 acres. Timothy grows best on clay loams and clay soils in the Great Lakes plain, the St. Lawrence and Champlain valleys and through the Hudson Valley.

In 1909 clover alone occupied 87,267 acres or only 8 per cent of that of timothy. Clover has declined on soils of all southern and eastern sections of the State. Clover succeeds best on a loam to heavy loam soil, moderately supplied with lime, with good drainage, such as in the counties of Genesee, Livingston, Oneida, Seneca and southern Cayuga, Onondaga, central Madison and northern Schoharie. The yield averages a little over a ton to the acre.

Alfalfa was introduced as early as 1790 under the

French name of lucerne. Robert Livingston, in 1793, had fifteen acres in Jefferson County. It was tried in the central part of the State in 1812 and while not eminently successful, some plants from that parentage are believed to have persisted for many years. The successful culture of the crop began in Onondaga County in 1867, with the introduction of

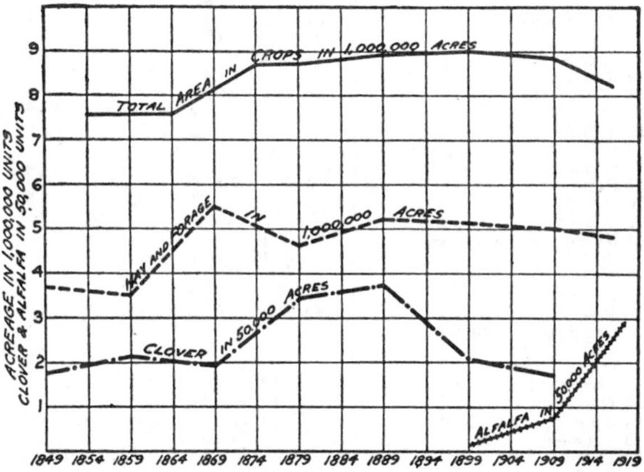

Fig. 20. Graphs showing by decades the total area in crops and the area in hay and forage, in clover and in alfalfa.

seed in chaff from the Pacific Coast. Doubtless, the early development of the crop was handicapped by its exacting requirements of a lime-rich soil and by the successful growth everywhere of red clover, and the greater ease with which clover can be handled in the rotation. The average yield of alfalfa is more than

TABLE IV. SUMMARY OF THE AREA IN CROPS IN ACRES IN NEW YORK SINCE 1844 *

YEAR HARVESTED	1844	1854	1864	1865	1874
1 Corn	595,134	917,601	632,213	721,600	627,749
2 Corn for silo....	....	....	....	....	....
3 Corn or sorghum grown for feed or for fodder only..	....	....	‡14,579	‡12,430	‡58,712
4 Oats	1,026,915	1,349,384	1,109,910	1,163,545	1,327,975
5 Barley	192,503	212,608	189,030	243,125	211,560
6 Buckwheat	255,495	293,233	240,302	184,717	263,781
7 Winter wheat.. } 8 Spring wheat.. }	1,013,665	{ 601,142	399,919	406,577	558,069
		{ 194,346	113,115	104,992	96,027
9 Rye	317,099	281,715	234,670	233,214	257,736
10 Flax	46,089	11,764	23,874	18,242	....
11 Field beans	16,231	16,918	61,821	24,208	81,321
12 Alfalfa .... } 13 Other hay .... }	....	3,384,441	4,237,085	4,296,721	4,796,739
14 Cabbage		....	....	....	....
15 Potatoes	255,762	220,576	235,058	254,403	358,433
16 Hops	....	9,482	24,339	25,870	28,278
17 Tobacco	....	786	12,981	4,614	2,593
18 Nursery stock and flowers	....	....	....	....	....
19 Root crops raised for stock food...	§15,322	§7,578	§8,124	§5,632	....
20 Peas	117,379	48,155	46,401	40,725	22,384
21 Canning factory crops	....	....	....	....	....
22 Other vegetables and garden	....	¶12,591	¶10,837	¶10,110	....
23 Misc. crops......	....	4	102	58	....
24 Apples	....	....	....	....	....
25 Peaches	....	....	....	....	....
26 Pears	....	....	....	....	....
27 Plums	....	....	....	....	....
28 Cherries	....	....	....	....	....
29 Quinces	....	....	....	....	....
30 Vineyards	....	....	....	....	....
31 Small-fruit	....	....	....	....	....
32 Total except hay and fruit	3,851,594	4,177,883	3,357,275	3,454,062	3,894,618
33 Total except fruit	....	7,562,324	7,594,360	7,750,783	8,691,357
34 Total	....	....	....	....	....

This table is adopted from the Census of Agricultural Resources of New York State for 1917.

* Statistics for 1879, 1889, 1899, 1909, are from the U. S. Census. All other figures are from the State census.

From 1844 to 1875, inclusive, for all crops except hay, the figures are acreage sown or planted. After 1875, for all crops except fruit the figures are acres harvested. For 1874 and all later years, the hay figures are acres harvested.

In 1875 the number of apple trees of all ages was reported. This number was divided by forty-five to get the approximate number of acres. The trees were then planted closer than is now the case.

	1875	1879	1889	1899	1909	1916	1917
1	673,284	779,272	493,320	658,652	512,442	341,931	327,509
2	....	....	....	†189,601	†259,119	365,990	321,926
3	‡49,929	....	....	....	....	....	112,511
4	1,438,166	1,261,171	1,417,371	1,329,753	1,302,508	1,120,565	1,083,646
5	282,393	356,629	349,311	111,658	79,956	95,189	102,297
6	201,894	291,228	280,029	289,862	286,276	263,667	269,138
7	564,221	736,611 (7–8)	462,561 (7–8)	557,736 (7–8)	285,823	351,225	351,594
8	69,836				3,266	12,536	26,821
9	229,070	244,923	236,874	177,416	130,540	118,176	111,458
10	....	....	2,922	159	58	....	....
11	78,641	....	....	129,298	115,698	198,309	218,742
12	4,783,857	4,644,452 (12–13)	5,243,010 (12–13)	5,582	35,343	163,150	146,221
13				4,959,782	4,748,249	4,140,489	3,970,379
14	....	....	....	25,261	35,269	41,668	47,244
15	405,806	340,536	357,464	395,640	394,319	309,213	348,269
16	37,004	39,072	36,670	27,532	12,023	....	4,516
17	2,516	4,937	8,629	11,307	4,109	....	2,095
18	....	....	....	9,734	11,659	....	6,258
19	....	....	....	....	663	....	18,032
20	23,645	....	....	14,748	4,007	....	....
21	....	....	....	119,130 (21–22)	140,246 (21–22)	44,661	43,608
22	....	....	....			64,858	84,624
23	....	....	1,188	3,384	3,763	30,678	....
24	406,192	....	....	....	351,918	342,243	385,737
25	....	....	....	....	29,213	50,050	53,470
26	....	....	....	....	22,777	37,187	41,031
27	....	....	....	....	7,796	8,733	10,249
28	....	....	....	....	6,356	12,959	12,989
29	....	....	....	....	1,936	....	3,277
30	....	....	....	....	59,340	54,020	56,126
31	....	....	....	25,051	22,496	28,841	32,893
32	4,056,399	4,054,379	3,646,339	4,050,871	3,581,743	3,358,666	3,480,188
33	8,840,256	8,698,831	8,889,349	9,016,235	8,365,335	7,662,305	7,596,788
34	....	....	....	....	8,867,167	8,196,338	8,192,560

In 1910 the census gave numbers of trees, but not acres. To get acres, apples are divided by 40, peaches, pears, plums, and cherries by 160, and grapes by 600. In the case of apples the figure used is an average found by orchard surveys in four counties. Since the census enumerated trees, it is probable that many were listed that were not reported by acres. The increase in the eight years 1910–1917 was therefore more than the figures indicate.

† "Coarse forage," mostly corn for the silo.

‡ Corn for fodder.

§ Turnips.

¶ Market gardens.

twice that of the other hay plants, 2½ tons to the acre, but 4 to 6 tons in three cuttings in a season is not uncommon.

Millet, or Hungarian grass, occupies a minor place as a home forage plant. It is a summer crop often put in where spring crops have failed. Its largest acreage is in the dairy counties of Delaware, Chenango, Cortland and Madison. The average yield is nearly a ton and a half to the acre.

Orchard-grass is the most common and the most widely distributed run-wild grass in the State. In June, its vigorous upstanding stems may be seen coming into bloom. Almost everywhere conditions seem favorable to it. Seldom is it planted in the regular grass mixture. Its tussock roots are vigorous and long-lived. It grows well with alfalfa and sometimes is used in seeding that crop, to fill in wet places unfavorable to the alfalfa. Cut at the same time as alfalfa, the stems are soft and palatable and make a good hay, but if left standing it soon becomes woody and makes poor hay.

In 1909 nearly 50,000 acres of grains were cut green for forage. Doubtless it was mostly used for soiling dairy cattle as there is an increasing tendency to use this system, together with the summer silo, and thereby reduce dependence on pasture. Pasture is a poor means of producing forage economically and outside of the use of waste land, many farmers are eliminating it. Rye, wheat, barley and oats form a good succession from early spring until corn is available.

Roots to the extent of 663 acres, probably beets,

were grown for forage. The chief objection to this crop is the hand labor required to grow it. The crop is rather exacting both in soil and culture. The soil must be rich and calcareous to give best results. The dry nutrients of beets have been shown to be equal in feeding value to grain concentrates. Beets are prized in feeding dairy cows for record milk yields because of the variety it gives the ration as well as its large feeding value.

Corn for silage is the preëminent green forage crop. Nearly 260,000 acres of coarse forage is produced and gives an average yield of seven and a half tons of green forage to the acre, equivalent to about two tons of dry matter to the acre. Thus corn stands next to alfalfa in its ability to produce nutrients. Corn, however, makes a good growth over a much larger area than does alfalfa. The area of corn for coarse forage is equal to half that devoted to corn for grain. New York is far from being a corn-producing state because of the cool summer climate, but for succulent forage, corn has no equal. A dairy farm without a silo is now regarded as unprogressive and relatively few of them exist. A silo of smaller diameter for summer use is now coming to take its place together with the larger winter silo in the adoption of the system of summer feeding. High summer rainfall, warm temperature, and a moderately calcareous soil combine to produce the largest yield of green-corn forage. The best yields are secured in the counties of Jefferson, Herkimer, Madison, Chenango, Delaware, Montgomery and Livingston, where the average is about nine tons. In other parts of the State it falls to six or seven tons.

PASTURE

Pastures occupy a large area, about one-fourth of the total farm land. This does not include woodlot pasture. Much land too rough and stony to cultivate has been cleared and retained in permanent pasture. In different parts of the State the proportion of pasture land varies in inverse ratio to the general development of the region. It is lowest in the counties directly south of Lake Ontario, medium in the southern tier of counties and high in eastern and northern New York. Naturally the mountainous regions have the most pasture as well as the most timber. It embraces steep mountainous country, wet areas and other rough stony land. It is not standardized as to vegetation, condition, or yield of forage. Receiving little care, it returns usually a small amount of crop. The stony pasture is often the earliest land to green up in spring because the soil is thin and easily warmed.

In the wet areas, red-top is probably the most common grass. Legumes do not thrive generally. The small white or Dutch clover comes in naturally on many sweet soils. It is closely associated with Kentucky blue-grass. Wild swamp grasses play a small part in the herbage. In all the regions of better soils, where lime rock is prevalent and the soil calcareous, Kentucky blue-grass prevails. All through central and western New York this is the common grass that not only overspreads the permanent pasture land but encroaches into the alfalfa and mixed hay meadows. Its presence and growth is a good index to the native character of the soil. In Wayne, Cayuga, Onondaga and Madison counties, and the southern slope of the Mohawk, there is much steep land that is otherwise good soil, being the calcareous glacial till laid into drumlin ridges or eroded into narrow deep valleys. These make admirable blue-grass pasture land. A little further south the limestone outcrops of the Niagara and particularly the Helderberg formations that have a thin pocketed soil covering, the Honeoye stony loam soil, are left in pasture. Up the St. Lawrence and in the Hudson Valley are smaller areas of calcareous pasture-soil associated with the Trenton-Calciferous limestones. The light sandy areas of the lake plains are often occupied by a poor quality of pasture but more often are in timber. The higher hills of southern and eastern New York are largely devoted to pasture, the cultivated crops occupying only about 10 per cent and all tame crops not over half of the area. In that region is much pasture of poor quality. Over large areas the soil is so depleted that Kentucky blue-grass does not occur and its place is taken in part by the smaller-growing Canada blue-grass (*Poa compressa*) which makes good feed but "small picking." In the last stage even the Canada blue-grass is crowded out and fails to maintain itself against the advances of wild strawberry, devil's paint brush, white daisy, red sorrel, and ferns that make up the bulk of the herbage. This transition in vegetation is closely identified with the increase in the acid condition of the soil and the general reduction in the content of organic matter. In the main, the higher the elevation the poorer the condition of pastures. These poor pastures are not confined to a single soil type, but follow certain groups as the Volusia-Lordstown, the Dutchess soils and the Gloucester and Culver series.

Since most of this land is difficult to plow, it is a problem how to keep the pasture in condition to produce even a little forage. The favorable conditions for the growth of bushes and trees cause a large variety to spring up and compete with the grass. Often the pastures are over-grazed in hot dry weather and no materials, either seed or fertilizer, are applied. The droppings of cattle are poorly distributed, and if the growth of desirable vegetation is stimulated it is poorly utilized by stock. That it is possible with some attention markedly to improve poor pastures has been demonstrated.

GRAIN CROPS (See Figs. 18, 19, 21, 22)

Corn for grain is most extensively grown in the region lying south of Lake Ontario. It is also a prominent crop in the valleys of southern and eastern New York on both the bench soils and on the first bottom land. Before settlement by white men, corn was extensively grown by the Indians in these latter positions.

On the hill land above an elevation of 1000 to

1200 feet, corn does not mature with regularity. This keeps it largely off the hills of southern New York. In New York, as in other parts of the country, corn is the safest tilled crop on overflow lands and the surest of the grains on very rich swamp lands, such as the

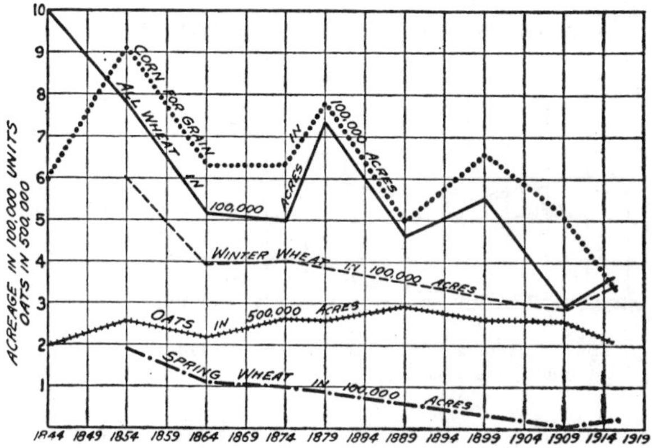

Fig. 21. Graphs showing by five-year periods the area in corn for grain, all wheat, spring wheat, winter wheat, and oats.

Clyde series and muck soils. Of the upland soils, the Ontario series is most widely used for the crop.

Because of the seasonal limitations, flint varieties are extensively grown on the hills and dent varieties prevail at the lower levels.

The State produces about a half million acres of corn, with a ten-year average yield of 36.2 bushels. The two tiers of counties south of Lake Ontario have three-fourths of this area. The average yield is also largest in that region. The three counties of Wyoming, Genesee and Monroe each have an average yield of 40 bushels or more to the acre. The northern and eastern counties fall considerably below the average yield.

There has been a continual decrease in the area

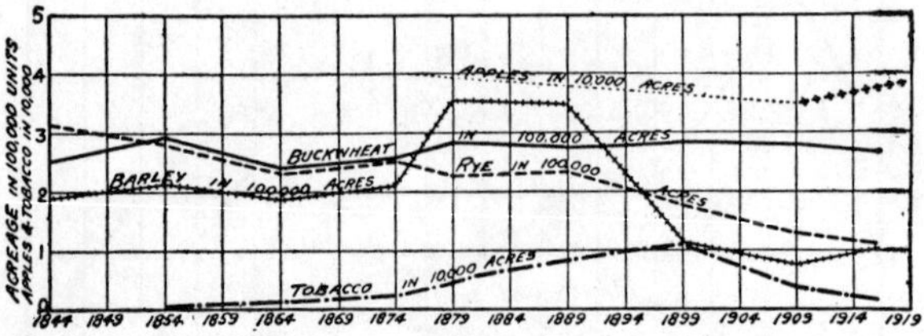

Fig. 22. Graphs showing by five-year periods the area in buckwheat, barley, rye, apples, and tobacco.

devoted to corn for grain in the last thirty years. In the census of 1880, 780,000 acres of corn were reported. The decrease has been most pronounced in those counties outside of the corn-belt of the State but has occurred in all parts. When combined with the area devoted to silage corn, the total is about the same as the maximum acreage the State grew in 1880.

Oats occupy nearly three times the acreage of corn and four times that of any other cereal. They are largely produced in all sections of the State. The acreage amounts to 1,300,000 and the ten-year average yield is 32.2 bushels or about half as much substance as an acre of corn. It is produced on all grades of soil and under a much wider range of climate than is corn. Its successful growth reaches further up on the hills and further north in the State than does corn. One reason why oats have a large place is that the seasons are a little short to start a fall crop such as wheat or rye after a summer tilled one. Oats may be put on early in the following spring. Another reason is its value as feed and when ground with corn it is a standard constituent of rations for horses, poultry, and to a considerable extent for cattle.

Oats grow best in a cool moist climate which is common in New York. They respond to the better soils and the best average yields, 30 bushels or more to the acre, occurred in 1909 in the same region in which the best yields of corn are secured. Monroe County produces 35, Ontario and Genesee 33, Livingston 31, Wyoming and Seneca 30 bushels. However, Chenango County, in the center of the State, produces 32 bushels and Franklin in the extreme north 30 bushels. The southern tier of counties have yields around 20 bushels or below, doubtless reflecting the less productive soils of the hills where it is one of the common grain crops. Oats are widely used as a nurse crop for seeding grass. The acreage of oats has held its own or slightly increased over that of 1880.

Wheat occupies about half the acreage of corn, around 300,000 acres. Its production is almost entirely confined to the two tiers of counties directly south of Lake Ontario. It is pushed back from the lake somewhat by the development of fruit-growing. The eastern half of the State produces less than 50,000 bushels a county and many of them only a few hundred bushels. The counties of Ontario, Livingston and Genesee normally have an average yield of nearly 25 bushels. The largest total production of wheat was in 1850 when it attained 13,100,000 bushels. The acreage was then about 750,000. Soon after that time, the "wheat weevil" appeared and caused such damage to the crop during the succeeding fifteen years as to force a change in the rotation and a considerable abandonment of wheat-growing. The acreage was as low as 289,000 acres in 1910, but increased again during the World War to 425,000 acres. The ten-year average yield has increased regularly from 14.1 bushels in 1879 to 19.3 in the seven-year period from 1906 to 1912. Spring wheat was early grown in the higher hill regions, then abandoned and recently has again been grown to the extent of nearly 30,000 acres.

Wheat production fits in with that of beans and canning-factory peas. The former crop in particular, which occupies a large acreage in this same region, is harvested and out of the way in good season for planting wheat. Beans leave a clean, firm seed-bed favorable for wheat and the residual fertilizer and the legume residue combine to make the succession a good one. In that same region, clover seeded with timothy is the most common hay mixture, and wheat makes a good nurse crop for these. Further, or perhaps basic to all these facts, is the prevalence of clay

Plate IV. A grain field in the Genesee Valley.

and heavy loam soils of calcareous nature and fairly smooth topography that permits the use of modern harvesting machinery. Fertilizer is generally used on wheat and, in fact, in seeding all the small grains, but not often on corn.

There was a time when New York was the granary of the country, the period immediately following the opening of the Erie Canal in 1825. From the earliest settlement it was the commonest cash crop because of the ease of its storage and transportation. It continued to be the leading grain crop until the substantial opening of the Central West when its production shrank to the present acreage and to the region where it is now most successfully produced. Here it fits into the prevailing system of farming by which wheat is able to hold its own against further western competition. If not in itself profitable, it is worth while because of the aid it gives to the conduct of the prevailing system of farming.

Both barley and rye are secondary grain crops of limited area and emergency use. Barley is produced in the same region as wheat, south of Lake Ontario. Its growth swings a little further east and north than wheat, being used in the dairy region east of Lake Ontario and in the northeastern part of the State. It is a short-season crop and often finds place on wet, " late " land where oats or other spring grains could not be planted. It also reaches south well up on the hill land of the Genesee Valley, doubtless because of its ability to grow in a cool climate. It is almost entirely used as feed. The acreage of barley is from 80,000 to 90,000 and the ten-year average yield is 26.6 bushels. There has been a tremendous decrease in the acreage of barley in the last thirty years. Together with wheat it once had a large place on the hill lands, reaching its highest point in acreage in 1860. In 1879 it occupied approximately 360,000 acres.. Northwestern states' competition forced its decadence.

Rye occupies more acreage than barley and instead of one has two centers of production. It is grown extensively in the Finger Lakes and Genesee Valley regions, and again in the central section of the Hudson Valley. In the former region it takes its place with wheat in the rotation, perhaps on rather poorer soils. In the latter, it is the preferred grain on thin soil, is used extensively as a nurse crop for grass seeding and, in addition, the straw has a high market value in the city for packing and certain manufacturing purposes. It is carefully harvested and threshed to preserve the straw without breaking. Much of it is cut with a dropper and bound by hand and some is even cut with a cradle in the old-fashioned way. The acreage is about 150,000 and fluctuates considerably from year to year. The ten-year average yield for the period ending in 1915 was 17.3 bushels, three bushels more than the ten-year average for any period since 1865.

Rye is widely planted as a cover- and green-manure crop. It makes a fair growth on rather poor soil. It grows late in the fall and reaches good size early in spring. The fall growth may be pastured and the spring growth turned under for manure. In the dairy region where the soiling system is used, rye is the first crop available in spring. It is now much recommended for seeding with hairy vetch as a green-manure crop. They grow together very well, the rye serving to hold up the trailing vetch, and the support promotes the growth of the latter. They are ready to be plowed under about the same time in spring and usually in season for planting a regular crop.

While New York stands at the head of the list in the production of buckwheat, the acreage of that crop is fourth among the cereals. Wheat and also oats and corn stand above. The competition for third place is fairly close with wheat, in some years, exceeding buckwheat. The acreage now approaches 280,000 but may fluctuate widely in different years. This is probably because it is an emergency crop and is put in where the land is wet, or where for other reasons the regular crop has failed. As a general average, the acreage has not changed materially in the last forty years. It has dropped a few thousand acres. Pennsylvania is a close second in acreage and in a few years has exceeded New York. The high tide of production in the country was in the sixties, but the statistics do not show that New York produced much more than at present. In several years through the intervening period, it has exceeded 300,000 acres.

The production is nearly all in the lower half of the State. The two main centers are the southwestern part and the cultivated territory centering on Albany. In the former, the most extensive area is through the Finger Lakes and particularly on the higher land on the Pennsylvania line, centering in Tioga County. However, the production is heavy all through western New York back from the lake shores. Its heavy production reaches well north along Cayuga Lake. Buckwheat is considerably grown in northern New York.

The soils with which the growth of buckwheat are most associated are those of the southern plateau, the Wooster, Lordstown and the Volusia. The high elevation and moist character of this soil coupled with the fine fibrous root system give buckwheat the advantage over other cereals.

The ten-year average yield of buckwheat for the period ending in 1915 was 20.4 bushels, being 2 to 7 bushels more than in any preceding ten-year period since 1870. The yield varies in different counties from 12 or 14 to 28 bushels, the last being in Cayuga County. Evidently it is very susceptible to differences in season. Madison yielded 26 bushels in 1909, Chautauqua 25.6, Steuben 14, Tioga 21, Albany 19.5, Rensselaer 21 and Clinton 19 bushels.

Buckwheat is perhaps most widely used as a cover- and green-manure crop, particularly in orchards where tillage is practiced. Planted in early summer, it will make a good carpet by apple harvesting time. If a hardy legume is added, this will fill-in the fall and following spring. In the hill country, the crop is often used as a nurse in seeding grass.

The large acreage of buckwheat is reflected in the heavy production of buckwheat flour that is ground in the smaller country mills. The crop is very practical for the State, the gross returns to the acre comparing very well with those from other cereals. It responds to good soil and on calcareous loams, such as the Ontario, the yield is much increased over the average.

VEGETABLES (See Figs. 18, 19, 23, 24)

Potatoes constitute the one staple crop in which there has been a consistent increase in acreage in the State for the last sixty-five years, as recorded in the decade census reports. Since 1850 the acreage has approximately doubled. It now ranges from 350,000 to 400,000 acres. While New York has been in the lead in the production of potatoes, Michigan is rapidly approaching in acreage, and in one recent year equaled New York, the yield, however, not being so high.

Potatoes are used for human food more than any other staple crop. The yields look large but it is an interesting fact that both corn and wheat produce more total nutriment and more protein to the acre. In starch production the potato exceeds wheat but still yields a third less than corn.

The relative production, taking the average yields for the State, are given in the following table:

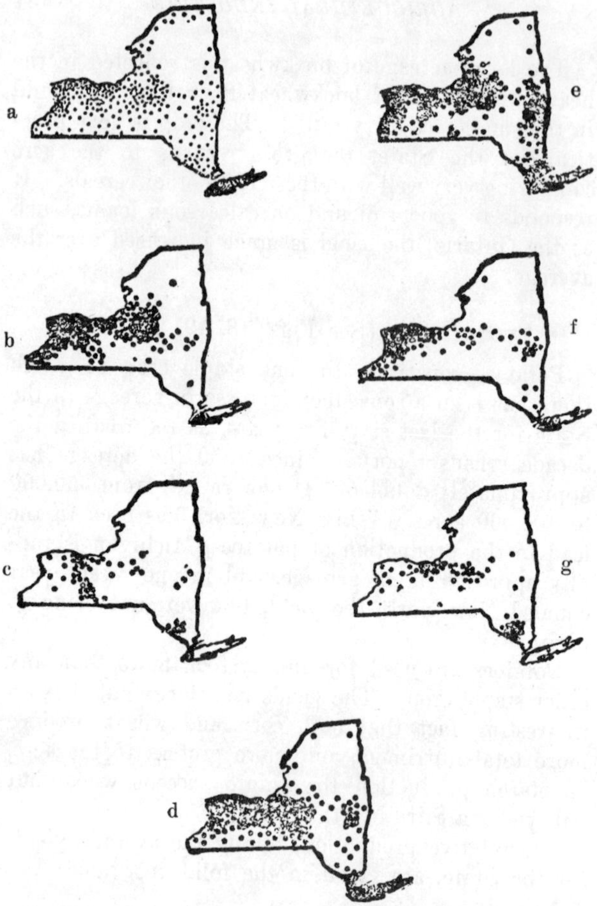

Fig. 23. Maps of the acreage and distribution of the specified vegetable crops in 1909. One dot equals 50 acres except for the total area of vegetables when one dot equals 500 acres. a. total vegetables; b. green peas; c. celery; d. cabbage; e. sweet corn; f. tomatoes; g. onions.

TABLE V. PRODUCTION OF POTATOES, CORN, WHEAT AND BUCKWHEAT IN NEW YORK

	Yield of dry matter, lbs.	Yield of protein, lbs.	Nitrogen free extract, lbs.
Potatoes, 100 bu.	1260	126	1036
Corn, 36 bu.	1950	220	1500
Wheat, 20 bu.	1080	142	870
Buckwheat, 20 bu. ...	845	96	625

New York has about 10 per cent of the acreage of potatoes and produces from 8 to 12 per cent of the crop grown in the United States.

By far the most important region of production of potatoes is the middle western part of the State from Lake Ontario at Rochester southward to the State line, and from Syracuse to the western divide of the Genesee River. The acreage thins out on either side of this region. Eastern Long Island is second in importance, a third and growing section being northern New York on the northeast flank of the Adirondacks. The upper part of the eastern Hudson Valley still retains a considerable acreage but it has fallen from its former large acreage.

The most marked change in the last fifty years has been the large increase in acreage in the western New York potato belt. In all the counties there has been an expansion of one-third to an entire doubling of the area. However, since 1905, there appears to be some diminution of the acreage in this territory and therefore in the State. The eastern counties had their largest production between 1860 and 1880.

The soils on which potatoes are grown vary considerably as to series but agree quite closely as to texture. They are all a sandy to silty loam. None is heavy clay or light sand. In the southern part of western New York, potatoes are grown on the hill lands on the Lordstown and Wooster loam and silt loam. The Volusia series in the same region is very much less suitable for the crop, probably because of its poor drainage. Both the former series are fairly

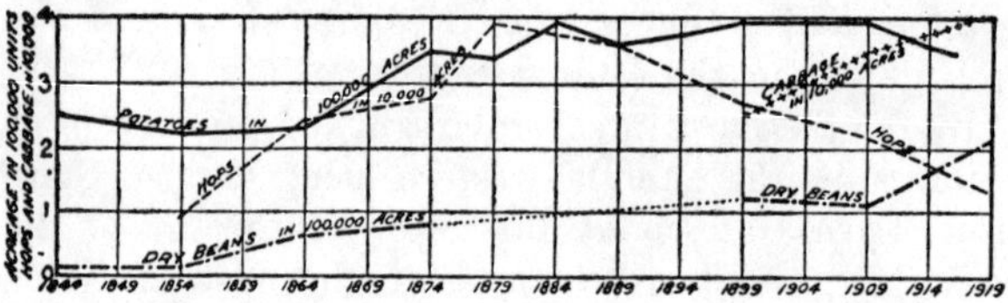

Fig. 24. Graphs showing by five-year periods the area in potatoes, hops, dry beans, and cabbage.

well drained, deep and well oxidized. Near the Ontario Lake shore, the Dunkirk and the Ontario fine sandy loams are most utilized. In western New York, late potatoes are a staple general farm crop in rotation and receive from four to eight hundred pounds of a complete fertilizer to the acre. The third center of production is in Clinton and Franklin counties, largely in the siliceous sandy and stony mountain soils of the Coloma series. Their best yield is on rather open fine sandy loam at an elevation of 1200 to 1800 feet. Often potatoes are the first crop on new land. Very little fertilizer is used. While grown in rotation, modern machinery is not employed.

The ten-year average yield of potatoes for the State for the period ending in 1915 was 97.5 bushels. It has ranged from 145 bushels in 1914 to 62 bushels in 1915. The northern counties have the best yields and run from 150 to 200 bushels. Next to these stands Long Island, then the western potato belt. The Hudson Valley counties have the smallest yield.

Both dry beans and green peas are important crops and are grown in the Finger Lakes region of western New York. In the production of beans New York is third, with an acreage of 116,000. Michigan and California lead with three-fourths of the total for the country, New York being the only other state with more than 100,000 acres. New York's acreage has been going down for forty years, the fall being heaviest in the counties of the Mohawk and the middle Hudson Valley where the crop was formerly of only minor importance. In recent years the growing of beans has been much hindered by a diseased condition of the roots that carries over in the soil and has caused a heavy reduction in acreage in the centers of bean-growing.

In green peas the State leads with a fifth of the total acreage, about 17,000 acres or one-seventh of that of beans. Wisconsin is a poor second. Like beans, green peas have suffered a heavy decrease in acreage that has been continuous for several decades and has been quite heavy in recent years.

These two legumes are associated with the calcareous soils. The Ontario and Honeoye series are their natural habitat. They run over onto the sandy loam and clay loam members of the associated Dunkirk soils. They are also grown to advantage on the alluvial soils in the region where they are dominant. A fairly sweet condition in the soil is a prerequisite to their successful growth.

Beans occupy the more restricted region and are grown from the west shore of Cayuga Lake nearly to the Niagara River. The Genesee River represents the central axis of the region. It spreads along Lake Ontario but does not reach the Pennsylvania line. The plateau region is not favorable to the crop, although it runs up on the Lordstown and Wooster soils along the southern boundary. Livingston and Monroe counties have the most extensive acreage.

Rainfall is a large factor in determining the growth of beans. In the region in question the harvesting period, September, is rather dry which aids in the field curing process. Red kidney, yellow eye, common white and pea beans are grown. The average yield of all varieties for the State is 15 bushels. The yields increase toward the dry region to the west, rather than toward the moist sections to the east and south.

Beans are grown in rotation, preferably on a clover sod. They are usually followed by wheat, as the requirements of the two crops dovetail very well. This rotation of beans, wheat and clover is one of the most effective in use in the State to build up a soil if the second growth of clover is turned under.

Sheep are largely kept in the bean-growing region and are used to consume the bean straw.

Green beans have a small development near the larger cities of Utica, Syracuse, Rochester and Buffalo.

Green peas have a wider spread than dry beans and reach well down the Mohawk Valley and west along the south shore of Lake Erie. They are even more sensitive to the acid soils of the southern tier counties than are beans. However, they are less sensitive to summer rainfall, since they are cut and shelled green. They, therefore, reach east into the region of high rainfall between Syracuse and Utica. This is also an important dairy and stock-raising country, and the green pea vines are ensilaged at the canning factories and sold back to the farmers as feed. Statistics on yield are not available, but observation indicates it to be 18 to 20 bushels to the acre. Green peas are grown in the rotation similar to beans but have a wider crop association. Peas are an excellent forerunner of fall grains and to precede alfalfa seeding.

Dry peas are grown in the same region as the green peas and occupy about the same acreage as green beans, about 4,000.

Cabbage production reaches from Syracuse to Buffalo and from Lake Ontario to the break of the higher hill land in the southern part of the State. It also reaches well into the valleys along the northern margin of the hill country, for example, in Cortland

County. Like beans and peas, it follows the calcareous soils but is rather less sensitive to a deficiency in lime. Moist climate and a nitrogen rich soil favor its growth as they do all vegetables. For largest yield it inclines toward the swampy soils such as the Clyde and muck but the greater part of the acreage is on upland soils of the Ontario series. The crop is stored and widely shipped. Near Syracuse a considerable amount of kraut is made. The area occupied by cabbage is about 35,000 acres which puts the State in the lead by far, with a fourth of the total crop in the country. The yield is from 8 to 12 tons but some farms exceed 20. Cabbage is a good cleaning crop for the land by reason of thorough tillage, and by many farmers is used to combat quack-grass.

Sweet corn for direct sale and for canning occupies about 24,000 acres in the canning-crops region south of the Great Lakes, and also in the Hudson Valley from which latter section it is shipped to New York as green corn. Its distribution is much the same as green peas.

Tomatoes are largely associated with sweet corn and green peas. They thrive on strong fertile soils and seem to succeed best in the region of low summer rainfall, closely fringing the edge of Lake Ontario and the south shore of Lake Erie. It is possible that the larger amount of sunshine and the dry atmosphere that prevails in this region promotes ripening and reduces susceptibility to disease of the fruit. The acreage is approximately 8,500 or about equal to the combined acreage of cabbage and onions.

Onions constitute a prominent crop and are most largely grown on muck soil, for that reason following the primary development of such soils through western New York from the Niagara River eastward to Rome in Oneida County. Wayne County leads although the South Lima area in Livingston County and the sections around Rome receive rather more popular advertising. In the lower part of the State in south central Orange County, large areas of muck soil are almost exclusively given over to the production of onions, so much so that "Orange County reds" have a standard market quotation as a recognized grade. Heavy fertilization strong in potash is required for all crops on muck soil. The area of onions is about 5,500 acres and New York ranks a close second to Ohio, several other states following closely. The yield averages 300 to 400 bushels to the acre, though two or three times this amount are secured on some farms.

Celery ranks with lettuce and onions as a muck-land crop. New York led in area of these in 1909 with about 3,000 acres or a fifth of the total in the United States. Muck areas in the Genesee Valley region are most largely devoted to celery, and the crop extends into the southern tier of counties much more than does onions. While the yield is larger on muck than on upland soils, and perhaps the quality a little better, it is recognized that the keeping properties in storage are better for the crop from the rich silty mineral soils, particularly dark alluvial soils, such as the Genesee and the Papakating series.

Other crops having a localized development are cauliflower, Brussels sprouts, Lima beans and cucumbers. On the medium sandy loam and in conjunction with the peculiar climatic conditions of the north shore of Long Island east of Wading River, these vegetables attain peculiar perfection and are most largely grown. The chief center of production of sprouts is at Orient, while that for cauliflower is a little to the west. The production of Lima beans is associated with cauliflower. While cucumbers are grown to a considerable extent near Orient, the larger area is near the western end of Long Island in western Suffolk County and nearest the south shore. The acreage is between 2500 and 3000 used both for immediate consumption and for pickling. Yields run as high as 150 barrels to the acre, but less than 100 is common. The yield of Brussels sprouts is about 2000 quarts and the area occupied is about 150 to 200 acres.

Mushrooms are grown by the acre in abandoned natural cement rock mines and in a smaller way in special cellars. Akron, in Erie County, is the largest center of the industry in the country. Here mines of large area a little below the surface of the ground have been abandoned for purposes of making natural cement. The beds are laid on the floor and the height of the ceiling permits the free use of wagons and teams. The total area of the mine is about twenty acres, of which six to eight are used for the crop.

The total area in vegetables exclusive of potatoes is about 175,000 acres. The crops listed above occupy about 100,000 acres, leaving 75,000 acres for minor crops and minor areas.

An important center of green vegetable production is along the shore of Lake Erie southwest of Buffalo. Owing, doubtless, to the general increase in population in the northeastern part of the United States, and to the markedly perishable character of many of the vegetables, the production of these crops is on the increase in this section of the country—about equal to the increase in population in cities in that region.

The sugar-beet fiasco of the late nineties and the early years of the present century is an excellent example of the futility of introducing a new crop without full knowledge of soil conditions adapted to its growth. Under the stimulus of state subsidies, factories for making beet-sugar were erected at Binghamton and at Lyons. The first factory was clearly headed for failure from the very start because the area of soil adapted to the culture of the crop in that region is confined to the river valley and is very limited in extent. At Lyons, similar difficulties existed but not of so serious a nature. Here, too, the soils are better suited to other staple crops than to beets. Had one of these factories been located in the region of Buffalo, where there are large areas of soil in Niagara and Erie counties adapted to its culture, it would have had some chance of success.

FRUIT (See Figs. 22, 25–27)

In acreage and value of fruit New York is second only to California, which has three times as large a total area. One acre in every fifteen in crops in New York is devoted to fruit-culture. The total area in all fruits is approximately 595,000 acres, equal to the total acreage of potatoes, vegetables and special crops, one-fourth that in cereals and about the same as that of corn. The value, $25,000,000, in 1909 was five-ninths that of the cereals and five-twelfths that of all the milk sold in the State. The rank of New York among the states is second in quantity and value. The value of the fruit crop in New York is as large as that of all the tropical and subtropical fruit produced in the country. The State is first in apples, fresh grapes, bush-fruits, pears and quinces, third in peaches and plums, seventh in cherries and eighth in strawberries. There is even more diversity in the conditions that govern the production of different fruits than for general farm crops, and they must, therefore, be considered by varieties.

The apple far outstrips all other fruits in acreage and value, having about half the value of all fruit. There were in 1910, 11,250,000 trees of bearing age, and 2,750,000 younger trees. In 1917 the total acreage in apples, both young and bearing trees, was nearly 386,000 acres. For decades New York has led in apple-growing but the production and marketing has not reached the high standard that is attained in the western boxed fruit and that is now beginning to characterize the crop in this State due to State and

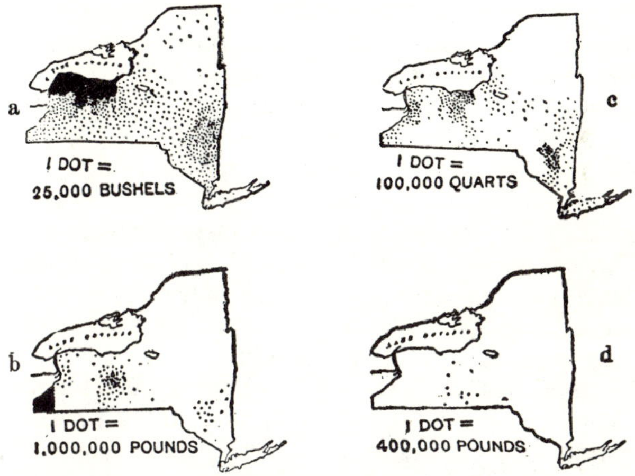

Fig. 25. Maps of the amounts and distribution of specified fruits and crops in 1909. a. orchard fruits; b. grapes; c. small-fruits; d. tobacco.

Federal laws requiring more systematic grading. The number of trees below bearing age, 20 per cent of the total, indicates the expansion to which the growers look.

The apple tree grows well in nearly every part of the State and the production of the fruit is widely distributed, but the commercial growth of the crop is restricted to well defined regions. Both soil and climate enter into the conditions that determine the distribution.

There are three or perhaps four sections of apple production: the lakes region, including the shore of Lake Ontario and the Central Lakes, the Hudson, and the Champlain Valley. The lakes region is by far the most extensive. Five counties on the Ontario shore, Niagara, Orleans, Monroe, Wayne and Ontario in 1909 produced more apples than any state except Pennsylvania. The lakes apple region forms a broad belt south of Lake Ontario and loops southward in the Seneca-Cayuga Lake district and in the Genesee Valley, to the middle of this part of the State. Ten counties stand in the following order in the number of trees to the square mile, a figure that very well represents the intensity of production: Niagara 1540, Orleans 1380, Wayne 1300, Monroe 1100, Genesee 620, Ontario 495, Yates 466, Erie 435, Seneca 370, Livingston 230. The acreage thins out to the eastward in western Oswego and Onondaga counties, and westward it follows the escarpment overlooking Lake Erie in Chautauqua County.

There are very few important orchards above an elevation of 1,000 feet and the more intensive production is at an elevation of 300 to 800 feet.

The Hudson Valley belt lies on either side of that river within ten miles, and centers about the north line of Dutchess County. It spreads north to Kinderhook and south to Newburg. It attains some importance in Saratoga County south of Ballston

Springs. The elevation here is below 500 feet.

The Champlain region is in the upper part of the valley back of Plattsburg in Clinton County at an elevation of 300 to 600 feet. There is also a smaller area of apple orchards along that lake in southern Essex County.

A growing period between killing frosts of one hundred and sixty days includes three-fourths of the

Fig. 26. Amount and distribution of specified fruits in 1910. One dot equals 500 acres. a. plums and prunes; b. grape vines.

commercial apple territory as it does of all fruits, and one hundred and fifty days marks the limit beyond which there are scarcely any large commercial orchards. This does not get outside of the ranges of elevation stated above, but neither does it include all of the territory within those limits.

The Indians in western New York had large plantings of apples when their strength was broken by General Sullivan in 1779. Apples were not grown for market in a commercial way much before 1825. The early plantings were based on the production of cider as well as fruit for culinary purposes. Beach states in "Apples of New York," that before 1840 probably not much over 15 per cent of the fruit was budded or grafted.

New York has contributed largely to the creation of new varieties. In a list of 804 given by Hedrick and his associates in Bulletin 361 of the Geneva Station, New York is credited with originating 144 varieties of those whose origin is known. Of these, eleven are adapted to growth widely in the State. Four are well-known commercial sorts, the Northern Spy, Jonathan, Newtown Pippin and Wagener. Four others belong to the Spy group. While the number of varieties is large, those of real commercial importance are small. Most of these appeared in mixed plantings of seeds carried into the new settlement by pioneers from the East. Probably at the head of these varieties is the Northern Spy. It appeared in a lot of seedlings grown by Mr. Chapin at East Bloomfield, where the site of the original tree has been marked. Early Joe and Norton's Melon appeared in the same lot of seedlings. Two varieties of Spitzenburg are in this list. The Wagener apple appeared at Penn Yan. The Yellow Newtown originated on Long Island, near Prince's nursery at Flushing. This is notable in another way because it was probably the first variety shipped abroad. In 1758 a package of this fruit was sent to Benjamin Franklin in London. In 1767, Robert Livingston sent a barrel of this fruit abroad. As late as 1825, apples were packed in straw-headed barrels. The Swaar apple appeared at Esopus in the Hudson Valley. The

Fig. 27. Maps of the acreage and distribution of the specified fruits according to age in 1909 except for apples which are for 1910. One dot equals 500 acres. a. apples (bearing); b. apples (below bearing); c. peaches and nectarines (bearing); d. peaches and nectarines (not bearing). One dot equals 100 acres. e. strawberries; f. bush-fruits.

Tompkins King originated in New Jersey but has been developed at Jacksonville in Tompkins County. The first white man's apple orchard west of the Genesee River was that of Schaffer on Indian Allen's farm at Scottsville, planted in 1799.

Beach in "Apples of New York" says that the Baldwin ranks preëminent above any other in importance in commercial orchards in the State. Rhode Island Greening ranks next in importance. Doubtless, these two varieties supply at least two-thirds of the apples grown for market. Neither of these varieties is a New York production. The Baldwin, named after Colonel Baldwin who recognized the merits of the fruit, originated in eastern Massachusetts. It is said by Wilson probably to have done more to give apple-growing a large commercial standing than any other variety. The Greening originated in Rhode Island in the vicinity of Newport. Next in importance is the Northern Spy. The relative rank of other varieties is not so easily determined.

The quality of New York apples is concededly high although the average color is not as strong as on western-grown fruit. The system of packing and marketing is improving rapidly under the influence of a set of State-established grades for apples and a system of bonding commission dealers. These laws are backed by the large membership of the New York State Horticultural Society. Marketing facilities, including grading and packing are improving and several central packing-houses have been established, notably in Niagara County.

In the apple districts, storage and manufacture facilities have developed to a very large extent to hold the crop to a longer marketing season and to utilize the low-grade fruit. Seventy-five per cent of the evaporated apples produced in the country are contributed by New York, mostly from the western district. To only a small extent the storage facilities are controlled by growers. The cold storage plants were formerly located in the larger market centers but the development of organization among producers is pressing home the suggestion that storage facilities near home that give a more broad and independent market outlook are to be preferred.

Peaches are a poor second to apples in value of fruit and number of trees. In 1909 their value was $2,000,000 or one-seventh that of the apple. There were 5,000,000 trees or about one-third as many as of apples. In 1917 there was reported 53,500 acres devoted to peaches of all ages. New York was eleventh in number of bearing trees, sixth in young trees, and third in value of peaches by the 1909 federal census.

The crop is much less cosmopolitan than apples and is confined to two districts, both of which are within the apple zones. The first is along the immediate shore of Lake Ontario from the Niagara River eastward and extending south but keeping rather close to the shores of Seneca and Cayuga lakes. The section adjacent to the Niagara River enjoys almost annual crops. The second district is in the Hudson Valley from Albany southward on either side of the river within five miles. The west shore opposite Poughkeepsie is most heavily planted.

The acreage has increased rapidly in the last ten years but had been nearly stationary for the preceding twenty years. The increase is shown by the proportion of trees below bearing age. In 1909 there were nearly as many trees below bearing age as were producing crops. Since the trees should live for twelve to fifteen years, this indicates a large increase. However, the very cold winter of 1917–18 killed many trees and correspondingly reduced the immediate prospects for large crops of peaches. The Elberta and the Early Crawford are the most satisfactory varieties for market, the former for the late season, the latter for the early season. In western New York, says Anderson, over half the plantings are of Elberta.

The peach crop experiences extremes in yield and market returns. No-crop years have been succeeded by seasons of over-production with slaughter prices.

The acreage and value of pears approaches that of peaches and in 1917 was reported at 41,000 acres. The regions of commercial production are the same as for peaches but pears reach out further from the influence of bodies of water, and as a rule are put on much heavier soils. The loam and clay loam types of the corresponding series used for peaches are employed for pears, and while important, lack of drainage is less critical than for peaches. The statistics appear to indicate a marked increase in acreage in very recent years. In 1909, the trees below bearing were equal to three-fourths those in bearing.

The preëminent variety for the State in acreage, value and successful production is the Bartlett. The Seckel and Bosc stand next in the list of commercial sorts. New York ranks first in the production of pears.

Plums and prunes are minor tree-fruits with less than a quarter the number of trees of pears and peaches. They also show a less proportion of young trees and the indications are that a decrease in acreage is imminent. Plums are closely associated with pears in distribution, climate and soil relations. They belong with the hardy fruits.

The cherry, unlike the plum, shows a strong tendency to increase in acreage. The acreage was somewhat larger than that of plums in 1917 and amounted to 13,000 acres and is increasing. While the State ranked ninth in number of bearing trees in 1910, it was second in number of trees below bearing age and fifth in value. The cherry compares with the apple in the breadth of its adaptation to New York conditions.

Cherry trees of native fruit of the cultivated sorts are common in a wild state and are found in every neglected fence corner and wayside spot. The cherry is the most common home-garden fruit throughout the State, the type being the pie or sour cherry which is more hardy and less exacting of soil and climate than are the sweet varieties.

The commercial development of cherry-growing clings rather closely to the main fruit belt in the Hudson Valley and the western lakes region. It has much the same spread as the pear and the plum. The sour cherries, according to Hedrick, make up 90 per cent of the plantings, the Montmorency, Early Richmond and English Morella leading in the order named. They are grown in the Hudson Valley on either bank of the river, but especially on the eastern shore in southern Columbia County, which has the largest number of trees of any county in the State. In this district the fruit is shipped to market as fresh fruit and the industry has not reached a high development.

In western New York there has been a very large increase in the acreage of sour cherries in recent years. The region of largest production is in the Seneca Lake district, but it is an important fruit through the Lake Ontario counties and has gained a considerable foothold on the Lake Erie shore in the grape district. The leading counties of the territory in the order of the number of trees to the square mile of area are Seneca, Monroe, Wayne, Ontario, Niagara, Orleans and Onondaga. Cherries, therefore, seem to show a preference for the eastern portion of this fruit belt.

The development in the industry of canning cherries has been chiefly responsible for the large increase in acreage. Geneva is an important center in this industry.

The sweet cherry, which is very fastidious of soil

and climate, is grown in only a small way and finds its best conditions on the lighter silty and fine sandy loam and gravelly loam soils, especially of the Dunkirk series from Rochester westward fairly near the Ontario shore, that is, within eight to ten miles.

In the production of quinces, New York leads in total value with Ohio and Pennsylvania carrying a similar acreage. There were nearly as many trees below bearing age as of producing age in 1909, a feature in which the State stood alone and indicating the increase in planting. At that time the total acreage was approximately 3000. The quince is uncommon outside the fruit belts but in the lakes region and Hudson Valley it is grown with peculiar success on the rich loam and fine sandy loam soils.

The prominence of New York in small-fruit production is due to raspberries which equal in acreage that of all other small-fruits. The first commercial planting was in Yates County in 1885. Raspberries grow best in the cooler regions, and among the northerly states New York leads in its production. The growing of bush-fruits is restricted to the area adjacent to the larger cities or to localities easily accessible to market. The raspberry is an exception and is most extensively grown rather remote from large cities, in a belt extending along the west side of Seneca Lake and thence northward into Wayne County. The three counties of Yates, Ontario and Wayne have half the acreage. Here the fruit is batted off the bushes into large baskets and it is then evaporated in dry houses. The soils of intermediate texture are adapted for its production and in this region the types are rather calcareous and belong to the Ontario and Dunkirk series. Whether the climatic conditions are peculiarly suited to the production of the fruit is not evident from the data at hand. Deficiency in soil moisture is particularly disastrous at the time of fruiting and there may be a connection between this fact and the moist summer climate that prevails in the Seneca Lake belt.

In the production of the other kinds of bush-fruits, New York has no especial rank. Their production has attained no large commercial importance. Currants, blackberries and gooseberries are grown with reasonable success and in amounts adequate to meet the market demands. With the possible exception of currants, the center of best production would appear to be in the states to the south of New York so that the increase in the acreage for consumption in the fresh condition beyond the normal growth of regional population does not seem to be advisable. In the Hudson Valley opposite Poughkeepsie and from thence south to Newburg, small-fruits are intensively grown and are largely shipped to New York by boat. Small-fruits are an important group in the lower Hudson region.

Strawberries rank next to raspberries in importance among the small-fruits but have only about half the acreage. Their distribution is fairly cosmopolitan in the valley and lake plain regions of the State. They are best grown on the silt and fine sandy loam types of series that are usually water laid, moist silt loam of the Dunkirk and Hudson series being perhaps preëminent. The common belief that the strawberry prefers an acid soil is hardly borne out by the distribution of its acreage. However, it is evident that a content of lime beyond what will insure ready decay of organic matter and, therefore, a high state of fertility is not essential. The commercial production of strawberries in Oswego County began about 1883.

It is interesting to note that 277 acres of cranberries on eighty-eight farms were reported in 1909. This was only about one one-hundredth of the total acreage. These gave a yield of 327,370 quarts worth $20,743. The acreage was double that of ten years before. This is all on Long Island, mostly on the eastern end. The production of the crop requires a rather high degree of skill. It is the only fruit that thrives in low wet soils and during the winter the crop is usually protected from low temperature by flooding the area with water so that the plants are submerged. The coast region of New Jersey supports half the total acreage of cranberries in the country which was about 30,000 acres.

Over 1200 farms report the production of nuts to the total amount of 2,750,000 pounds. The greater part were from black walnut, butternut, chestnut and hickory nuts presumably from trees growing wild. Eighty-one farms reported 456 English walnut trees in bearing that produced nearly 10,000 pounds of nuts. This tree is attracting considerable attention and orchards have been started in the lakes fruit belt. Scattered trees of bearing age occur at a number of points in the western part of the State. These give a fairly consistent yield of nuts. As yet, the propagation of trees from cuttings has been difficult and the bearing trees, which are seedlings, are of variable quality.

From almost the first settlement by the Dutch, attempts were made to grow grapes and make wine. The early efforts were with foreign varieties and these were all failures in the New World. All the grapes now grown commercially, except those used for raisins, have been developments from the native American sorts. The American varieties do not contain enough sugar and solids for making raisins. Richards of Manhattan Island had vineyards there and on Long Island and as early as 1664 enjoyed a monopoly in wine-making in that region. Robert Underhill at Croton Point on the Hudson, acting under the suggestion of Mr. Parmenter of Brooklyn who had given much study to grape-growing, developed the first important vineyard in the Hudson Valley. He began in 1827 with the varieties Catawba and Isabella, and expanded his vineyard to seventy-five acres. These same varieties also constituted the first plantings in western New York. The variety Isabella appeared as a mutant in the garden of Mrs. Isabella Gibbs in Brooklyn. In 1818 Deacon Elijah Fay planted the first grapes in Chautauqua County, near Brocton, which led to the foundation of the industry in what is now the Grape Belt. As late as 1859 there was not over 100 acres of vineyards in that region. In the Keuka Lake district the first plantings were in 1830 in a garden at Hammondsport by Rev. William Bostwick. The origination of the Concord variety by Bull in Massachusetts before 1850, which variety now occupies upwards of 90 per cent of the entire commercial grape area, was the real foundation of the development of the vine in western New York. Two well-known varieties, besides many minor ones, have appeared in the State. The Worden was originated by Schuyler Worden, of Minetto, Oswego County, in 1863, from seed of Concord. The Niagara grape, the leading American green variety, was originated at Lockport by Hoag and Clark in 1868, also from seed of Concord fertilized by Cassaday. The making of wine was the first outlet for the larger vineyards. In 1880 the first carload of table grapes from western New York was sent to Philadelphia by Jonas Martin, of Brocton. The commercial production of unfermented grape juice began in Westfield about 1900 and has had a rapid rise. California leads in grape-growing, her product being most largely the raisin grapes of European varieties. New York, which ranks second in total production, leads all other states in American varieties used for table and for making wine and unfermented juice. In 1910, New York was growing nearly 36,000,000 vines, covering about 52,000 acres. All but one-ninth of this number was of bearing age. The total number of vines was a little more than one-third of all the vines in the country outside of California.

The region bordering Lake Erie on the south, through the states of Ohio, Pennsylvania and New York, is the most prominent eastern grape-growing district. A large part of the acreage is in New York and forms the most important center of production in the State. This is the Chautauqua district. Three other districts of grape production are of less importance, in the order of acreage, being the Central Lakes, the Hudson Valley and the Niagara.

The Chautauqua district had its rise to commercial importance with the introduction of the Concord about 1860. It is located on the low and rather flat Lake Erie plain which is marked off from the southern New York highlands by a high steep slope. Grape-growing extends well up this slope toward the crest but has its best development on the old gravelly bars and beaches that stretch east and west near the foot. This area was formerly covered by lake waters. They are often calcareous in the subsoil. They are mostly of the Dunkirk series. Along the face of the slope the Lordstown and Wooster soils are dominant. The grape region is from three to four miles in width and extends westward from just within the Erie County line through all northern Chautauqua County and thence westward into Pennsylvania and Ohio, with ever decreasing intensity from the center of production around Brocton.

The Central Lakes region is made up of several

centers of production, the largest of which is the Keuka district. Next in importance is the territory extending southward from the lower end of Canandaigua Lake. The third section is more scattered and lies on either side of the middle part of Seneca Lake, reaching over to Romulus. These several districts are not connected by grape plantings. The land consists of the steep slopes along the deep valleys, the slopes being most pronounced in the first two districts. Topography and elevation appear to bring about the conditions favorable for the grape in this region. The Keuka-Canandaigua districts have a longer and warmer season than the Chautauqua and fruit matures two weeks earlier and is largely out of the way before the latter crop comes into market. The rainfall is higher and the air more moist, as a result of which the vines are more subject to mildew than in the Chautauqua region. The soils of the Seneca district are much the same in series as in the Chautauqua belt, Dunkirk, but the crop is grown more on heavy soils of the Ontario series. In the other two regions the soils are of the Lordstown and Wooster series, probably mostly of the former of silt loam texture and very stony. These latter formations are relatively thin and friable, which, together with the lay of the slopes, doubtless combines to give the large supply of heat favorable to the crop. The best parts of the Central Lakes districts are planted solid to grapes to the exclusion of other crops.

The Hudson district has been designated as the birthplace of American viticulture. Here are the oldest winery, the oldest vineyard, the first distributing point and the greatest number of varieties. Many of these have been originated there and some of them have been the basis of the best American sorts. Finally, there has been centered there a corps of viticulturalists whose names will long be perpetuated by horticulturists because of their many and varied contributions to the industry, chief among whom were the several generations of the family of Downings whose nurseries were at Newburg.

Like other fruits grown in this region, the grape follows closely the shores of the river from Newburg northward. It is a narrow irregular tract of stony mountainous country, with pockets of stony, sandy and silty soils of the Hudson and Dutchess series. Above Poughkeepsie and reaching across the river into lower Columbia County, steep but tillable slopes are commonly planted to grapes. The east slope is usually the warmer and on account of the direction of the prevailing westerly winds most protected from the wind.

The fourth or Niagara district lies in the angle between the Niagara River and the steep bluff overlooking Lake Ontario. It does not reach much east of Lockport and follows rather closely the foot and lower slope of this mountainous escarpment, on the stony and gravelly soils of the old lake beach. In these respects it is similar to the Chautauqua region. This district is far less prominent than the other three in extent of grape production.

The number of bearing vines in the several dis-

Plate V. Intensive fruit culture in the fruit belt in Western New York.

tricts in 1910 were: Chautauqua 17,100,000, Central Lakes 11,000,000, Hudson Valley 2,500,000 and Niagara 500,000. With the exception of the Chautauqua district, there has been a decrease in number of vines; here there has been a substantial increase but not in yield of grapes to correspond. It has been suggested by Morrison that the increased planting on unfavorable soils is responsible for this decrease in average yield.

In the Chautauqua district, the distribution of other varieties of grape than Concord is reported by Hedrick to be: Niagara 3 per cent, Wordens 2 per cent, Moors Early and Catawba each 1 per cent, other varieties, chiefly Delawares, 3 per cent. The only marked variation is in the Niagara district where the variety of the same name is grown almost exclusively.

The utilization of the grape has changed considerably and varies with the different districts. In the pioneer days nearly all the fruit was used for making wine. Table use of grapes increased until in 1890 it is stated that four-fifths of the product of New York and Pennsylvania were so consumed. Nearly all the Hudson and Niagara crops are still so utilized. The production of wine has never been an especially large factor in the grape-growing industry. However, the manufacture of sparkling wine or champagne has been important in the Keuka Lake section where 75 per cent of the product is made. This has been called the American champagne district and is on the same isothermal line as the famous Champagne region of France. L. J. Vance, editor of the *American Wine Press,* concludes that in recent years one-fourth of the grapes in the State were used for the production of wine and juice. However, wine has been produced commercially in all of the grape-growing centers.

The production of unfermented juice is increasing. In the Chautauqua region it is made entirely from the blue Concord variety. More than 3,000,000 gallons of juice are manufactured annually there, and production is increasing.

NURSERIES AND FLOWERS

In association with the large production of fruit in the State, there has been from early years a correspondingly large development of the nursery business. The oldest and most famous of these was the Prince Nursery at Flushing, Long Island, established in 1725. Four men of as many generations of the Prince family were proprietors of this nursery in succession, and were for many years leading authorities on important groups of American fruits and flowers, and authors of noteworthy catalogues and treatises, including one on the vine. Their nursery is said to represent the beginning of commercial fruit-culture in America. In 1793, their nursery and large collection of varieties became the Linnean Botanical Garden. It has now been long disbanded.

The Downing Nurseries, established at Newburg and supervised by several members of the Downing family, have occupied an important place in horticultural annals similar to those of the Princes on Long Island. In 1841, Andrew Jackson Downing, who is said to have been the first great American landscape-gardener, published his book on "The Theory of Landscape Gardening and Cottage Residences." Three generations of the Downing family of Newburg were notably prominent in the nursery business and in the development of fruit-culture.

The Rochester-Geneva-Dansville nursery district in western New York, probably the largest in the country, had its primary beginning when Patrick Barry went from the Prince Nursery on Long Island to Rochester in 1840. Of course, there were nurseries in that western district before the coming of Barry. Lincoln Fay, second son of Elijah Fay, had the first grape nursery in the present grape belt at Fredonia, another important nursery center, especially for vines and small-fruits.

Monroe County is now the leading center of the nursery industry in acreage. Rochester is the center of production. In that region occur large areas of fine sandy and silt loam soil favorable for the business, which naturally joins with the large fruit interest, of which that city is also the center. There are outlying centers eastward in southern Wayne County and in Onondaga County, westward as far as Lockport, southeast at Geneva and prominent areas at Dansville and Batavia. Dansville in the Canaseraga Valley in southern Livingston County is the largest shipping center. The Dansville nurseries were established by persons from Wyoming County, Pennsylvania, who began planting apple seeds about 1796.

The Dunkirk sandy and fine sandy, and especially the silt loam and the Ontario fine sandy loam are used for nursery purposes, except at Dansville where the alluvial Genesee silt loam is employed. At this point, tremendous storage and shipping facilities have been developed.

Flowers and plants are grown more intensively than any other crop. In New York the average acre value for the main groups of farm products is as follows: Farm forests $3.50, cereals $16.50, fruits $35.00, vegetables $51.00, nursery $320.00, flowers and plants $1725.00. These average figures appear disproportionate. It must be remembered that in most lines of production the average is far below what a progressive grower would call fair returns. This is due to the large number of farms that are unsuited to the crop or are poorly managed. The lower the acre value of the crop and the more extensive the nature of its production, the more does this disproportion between average and good returns appear. Doubtless, some such law is responsible for the remarkably good showing of flowers and plants, a large part of which are grown in glass houses.

The production of flowers and plants lies close to the larger cities. It represents a heavy capitalization in equipment and labor. As the center of large cities, New York has the greatest acreage which totaled 2,979 acres and was valued at $2,750,000. The acreage doubled in the ten years preceding 1909.

The glass-house farms for flowers are supplemented by open field areas for the propagation of certain types of plants. The industry is subdivided along the lines of types of flowers produced as the other forms of crop production are subdivided. Roses, carnations, gladiolas, peonies, violets, orchids, and others become the particular sphere of individual growers.

SPECIAL CROPS

Hops were once widely grown and important. For the last twenty years the production has been decreasing, due to the development of disease, especially mildew, and to competition from other parts of the country, particularly the Pacific Coast states. From the beginning of the growth of hops in Madison County in 1816, when James D. Cooledge marketed a bale in New York City, the area expanded until in 1880 it amounted to nearly 40,000 acres. Since that date it has been decreasing, first slowly but in the last fifteen years at a rapid rate until now it is less than a fifth of what it was formerly. In 1909, 12,000 acres were reported. In the height of their production hops were grown throughout the middle of the State and in the extreme northern part. The chief area of production has always been along the southern side of the Mohawk Valley. Here the hop dry houses, frame structures with a wind-vein ventilating hood on the roof, most of them now abandoned, are a conspicuous feature of the farm building group. The other important center of production was in the northern part of Franklin County, around Malone. A small amount has been grown in Livingston and northern Steuben counties. (See Figs. 19, 24.)

The distribution of hops and particularly their successful growth has been restricted to calcareous soils and a moderately high summer rainfall. The Ontario and Honeoye series, and to some extent the Mohawk soils have carried its chief development. The yield is from 800 to 1500 pounds, the average being near the lower figure. In Oregon, the leading state in the production of hops, the yield averages 1000 pounds or more and in California it is over 1400 pounds.

Teasel is better known as a common roadside weed in the northeastern quarter of the country. By transfer from England by William Snook a different variety became a farm crop in western Onondaga County in 1840. The town of Skaneateles is the center of its production and it is distributed over an area of perhaps a hundred square miles. The calcareous Ontario and Dunkirk soils of that region are used for teasel production. The crop is grown for the seed cones, the delicate hooked ends of the scales of which are used to raise the nap on cloth. A good yield is 100,000 cones to the acre. Like other special crops, this one is decreasing in acreage. A mechanical appliance for raising the knap on cloth has been devised with which the teasel cones must compete.

Tobacco is a commercial crop in two regions,— first a territory twenty-five miles in diameter centering in northwestern Onondaga County, and second in the Susquehanna-Chemung River valleys from Binghamton to west of Elmira. These regions had a total acreage in 1909 of 4,109 acres. Since that date the production has decreased considerably especially in the southern district. This is merely part of a decrease in the acreage of the crop that has been going on for forty years. It was formerly grown much more widely than at present and many counties through the lower elevations in the State reported an appreciable acreage in earlier years. Dutchess County still produced thirty-one acres in 1909. Tobacco seems to have been grown by the Indians. It became a commercial crop northwest of Syracuse about 1845. (See Figs. 22, 25.)

The soils upon which tobacco is produced are the gravelly, sandy, fine sandy loam and silt types of the Dunkirk series in the Syracuse district, and on the corresponding textures of the Chenango series in the Elmira section. The yield is from 800 to 1200 pounds to the acre. The production of the crop in New York State has never reached the high state of development that has been attained in other tobacco-producing centers, such as the Sumatra wrapper region near Quincey, Florida, and in the middle Connecticut Valley, or the cigar binder regions of Miami County, Ohio, and Janesville, Wisconsin.

Ginseng is grown in small areas in several regions, particularly through the middle of the State south of Syracuse. The most important localities are at Rose Hill and Scott, between Skaneateles and

Cortland. Its production is a very intensive business since an acre of mature roots well developed may be worth $10,000. The dried roots sell at $8 to $12 a pound. It is a medicinal root used by Orientals and requires three to five years for the roots to reach marketable size. A silty to fine sandy loam, moist and cool and moderately fertile, gives best results. An excess of lime carbonate promotes destructive diseases. While the soil must be moist, good drainage is essential for rapid vegetative growth. The Lordstown and Chenango silt loams are excellent types of soil for the crop. New York leads in its production.

Willows of the osier type for basket-making are a farm crop at several points in the western part of the State. The most important center is in southern Wayne County, where it was brought in by Dutch settlers. The crop is grown on moist, rich and moderately heavy soils. Lake and alluvial soils of the series common to that region are used.

Peppermint for its oil was formerly cultivated on the muck and swamp soils of Wayne County where onions and cabbage now find a large place. At one time that region was the center of the world's production, but it has now been transferred to the Clyde marshes of Allegan County, Michigan.

The sugar bush or hard maple orchard is still an important factor in the receipts of many New York farms. The tree is one of the most representative and cosmopolitan in the State, being common in every county outside the heart of the Adirondack

Mountains and the thickly settled and thoroughly cultivated districts. It does not thrive on thin, light or very sour soils, but eliminating these extremes it seems to be very tolerant of variations in soil. It is considered to indicate strong durable soil.

Over twenty-five thousand farms, or nearly 12 per cent of the total number, reported a total of nearly 5,000,000 maple trees in 1909. Vermont has for some time led in the production of maple sugar and Ohio in sirup, but in the total of both, New York ranks first.

The largest producing counties are those of the southern plateau region. It appears to be a marginal industry that stands midway between intensive farm practice and abandonment to straight forest purposes. The sugar grove is most common where the farm woodlot is a large factor on the farm, where there is a considerable amount of waste land, not steep and rough, but stony and difficult to till or remote from the homestead.

CHAPTER VI

ANIMAL INDUSTRIES OF NEW YORK

THE first Dutch settlers brought over many kinds of live-stock. In 1623, Peter Earstin Heft introduced the first cattle into New York, and in 1625 three boat-loads of stock were imported to New Amsterdam and included horses, cattle, sheep, swine, also seeds and plows. The presumption is that the cattle carried the Friesian blood. In those early days, cattle were important for draft purposes, as well as for meat and milk. As late as 1850 there was nearly half as many oxen on farms as horses. Horses were very much of a luxury among the pioneers when the ox could plow the field, draw the wagon, furnish meat, and finally provide leather for the shoes and boots for the household. Sheep were important to supply the wool that was carded and spun in the home into yarn, from which the winter clothing was largely made.

The animal industries have for many years held a leading place among the agricultural activities of the State. In the last sixty-five years, there has been very little change in the total intensity of animal production. Excluding poultry, in which there has been a large increase, the number of mature animal units is very little larger now than it was in 1850.

The ratio of animal units to the total number of acres in farms and to the total area of improved land has been remarkably constant. This ratio was 6.7 acres to the animal in 1850 and 6.55 acres to the animal in 1915. In 1880 it was 7.2 acres and in 1900, 6.3 acres. The total number of animal units for successive periods, excluding poultry, is shown in the following table:

TABLE VI.—NUMBER OF ANIMAL UNITS EXCLUSIVE OF POULTRY ON FARMS FROM 1850 TO 1915

Year	Number of live-stock units	Acres in farms	Number of acres per animal unit
1850	2,840,000	19,115,000	6.7
1860	2,910,000	20,975,000	7.2
1870	2,933,000	22,191,000	7.5
1880	3,305,000	23,781,000	7.2
1890	3,194,000	21,961,000	6.9
1900	3,607,000	22,648,000	6.3
1910	3,280,000	22,030,000	6.7
1915	3,368,000	22,000,000	6.5

In these figures, seven sheep and five hogs are estimated to be equal to one mature cow or horse. No allowance is made for young horses and cattle which are assumed to maintain a constant ratio to the total number in that class.

Some fluctuation is observable in the total number of animal units. The census year of 1900, for example, is a high point and at that time the number of acres to an animal, 6.3, was the lowest it has been at any time in the period under consideration.

This fact when studied in connection with the general supply and price of crops illustrates a general law to which the production of live-stock roughly conforms. Live-stock constitutes a reserve food supply, being a means of storing food and of converting it into new forms that increase the total consumption. Consequently, when the supply of food crops is large and the price relatively low, the total number of animals is increased and more crops are fed.

More persons can be maintained by the direct consumption of crops than on the animal products from the same amount of crops. In other words, live-stock are destructive of food values whenever those foods could be consumed by man. It has been found by investigation that 100 pounds of nutriment when fed to animals yields 15.6 pounds of nutriment in pork, 2.8 pounds in beef, and 2.6 pounds in mutton, while in milk it will yield 18 pounds, cheese 9.4, butter 5.4, and in eggs 5.1 pounds. On the other hand, when the crops are of such a nature that they are unsuited to human consumption but can be eaten by animals, they are thereby converted into food and other products suitable for human consumption.

The persistence with which live-stock is kept on the farms of New York indicates that there is some fundamental reason underlying the practice. It has already been pointed out in the chapter on crops that New York is not eminent in the production of grain crops that are used as food for animals. Compared with the middle western states, fruits and vegetables have a much larger place in the census figures of New York than corn and oats and wheat. It might be expected that the number of live-stock would respond more closely, especially since the region of large grain production in the United States is the great Mississippi Valley which is remote from the large eastern cities to which animals are more easily shipped than the crops on which they are fed.

The live-stock intensity is higher in New York than in the United States as a whole. It is 6.7 acres to an animal unit compared with 8.2 acres in the United States in 1910 and this is only slightly below the intensity in such an eminent producing state as Illinois, and is above that in the great corn-growing state of Kansas. The number of acres in farms to an animal unit in several of the middle western states is as follows: Ohio, 6.1, Wisconsin, 5.5, Kansas, 8.7, Illinois, 6.4, Iowa, 4.4, and Montana, 6.0

The crops on which the rather large number of live-stock in New York are chiefly dependent are hay and pasture. Nearly one-half or 9,800,000 acres of the area in farms is devoted to those crops. Nearly one-half of that area or 4,100,000 acres is in hay and the remainder or nearly 5,700,000 in pasture. Woodland pasture is here estimated at one-half value. Five-eighths of the pasture area or 60 per cent is of a sort that cannot be tilled. New York has a large area of land that is best suited to the production of a low grade of forage that can be utilized by live-stock and that would otherwise be unsuited for human use. The adaptation of the State to the production of hay, the capacity to grow corn for silage for winter feeding, and enough grain to make at least a beginning on the concentrate problem, together with the heavy demand for dairy products by the adjacent large cities, give the key to the prominence of livestock.

The pastures are notably good, due to the high summer rainfall. This is reflected in the cost of pasture an animal. G. F. Warren has collected figures showing that New York has the cheapest pasture land of the United States with the exception of the South. In all those regions where pasture land is abundant, the middle and southern parts of the State, the cost for pasture ranges from fifty cents to one dollar a month to an animal unit. The comparative figures as collected by Warren in 1910 are as follows:

TABLE VII.—COST OF PASTURE

	Cost of pasture a month
New York	$0.50
North Atlantic States	0.90
Corn Belt	1.35
Middle West outside of the Corn Belt	1.05
Blue Grass region	1.55
Southern States	0.30
Arid region	0.50
Western irrigated or moist land	0.65

Of the concentrated feeds used, a large part is purchased from outside the State. In the exclusively dairy region of Delaware County, a careful record on one hundred and fifty-seven representative farms for two years indicated that as high as 98 per cent of the grain is purchased. This is undoubtedly considerably higher than the average. The general effort on the part of farmers is to grow more protein in the form of leguminous hays, and to some extent roots, thereby reducing the amount of grain that must be purchased.

While the total number of animal units has been fairly constant, there has been a large shift in the relative number of the different types of live-stock. Cows have regularly increased in number throughout the last sixty-seven years. Other cattle have fluctuated considerably. Horses reached a maximum in 1890 and have been decreasing since that time with the exception of the last two or three years when there seems to have been some return of interest in their production. This may be due, however, to the increase in the acreage of tilled crops which require more horse power. Sheep reached their maximum in 1845 with a secondary maximum in 1900 since which date they have rapidly decreased until in 1917 their number, about 600,000, was only 10 per cent of that in 1845 and 35 per cent of that in 1900. Swine have decreased irregularly from their high point in numbers in 1840. At that date there was nearly 2,000,000 swine of all ages in the State. They had fallen off somewhat by 1860 and heavily by 1870, the post war period, after which they increased to 1890 when the number was 843,000. Since that year they have undergone a general decrease. In 1910 there were 660,000 swine, while in 1917 there were

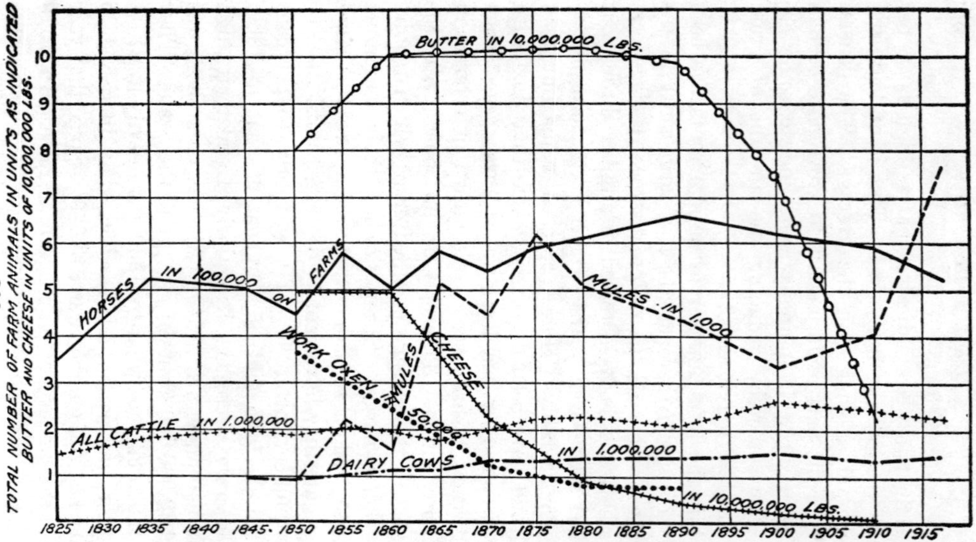

Fig. 28. Graphs showing by five-year periods the number of all cattle, dairy cows, work oxen, horses and mules and the production of butter and cheese.

less than 500,000. The decrease in the number of hogs is largely if not quite offset by the increase in poultry, the number of which has been growing regularly since 1880. (See Figs. 28, 29.)

It is to be expected that both the numbers and the prevailing types of live-stock in New York, located in the midst of the largest population on the American continent, should very accurately reflect the adjustment between such elements as the demand for perishable animal products, the use of waste foods of the farm and city, the relative supply of animal foods in the country, and the crops New York can produce to advantage. This adjustment, together with the relative efficiency of the different animals in utilizing foods, especially rough and waste foods, must constantly be kept in mind in studying the live-stock industries. Stock of all kinds are to a large extent scavengers, using the crops and waste material that man could not consume. This is the economic basis on which they must be handled, not only as to food but also as to labor.

In one class of animals only does New York stand first in numbers. This is in dairy cows, of which in 1910 there were 1,509,594; in 1915 there were 1,539,000. Wisconsin and Iowa are in second and third places respectively. In the total value of all live-stock, New York ranks eighth. The State is also eighth in the value of all cattle but ninth in total number.

New York is fourteenth in the number of horses. In number of sheep and swine, the rank is so low as to merit very little consideration in the total product of the country. In number of poultry, the State ranked tenth, and in eggs produced, eighth, indicating that more attention is devoted to egg

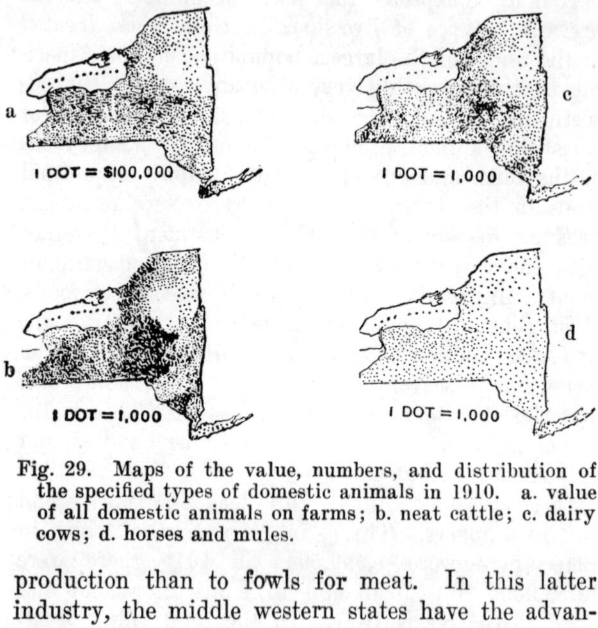

Fig. 29. Maps of the value, numbers, and distribution of the specified types of domestic animals in 1910. a. value of all domestic animals on farms; b. neat cattle; c. dairy cows; d. horses and mules.

production than to fowls for meat. In this latter industry, the middle western states have the advantage. In the number of colonies of bees, the rank is eighth. Corresponding to the large number of dairy cows, the State ranks first in the amount of milk produced and in its total value. On the other hand, in the products made from milk, butter and cheese,

New York has no particular prominence. It is twenty-second in the production of butter and fifth in cheese. This illustrates the demand by the large cities for raw milk which prevents its use for manufacturing purposes. (See chapter on manufactures.)

ANIMAL HUSBANDRY AND SOIL FERTILITY

In addition to the value of animals and animal products as human food and equipment and their use as conservators of otherwise waste human food material, still another viewpoint is frequently advanced in favor of live-stock husbandry. Animals produce manure as a by-product and in every intensively farmed region this is assigned high value to maintain and improve the productive capacity of the soil.

The three and a quarter million animal units in the State will produce manure at the rate of approximately 38,000,000 gross tons, or 7,500,000 tons of dry matter in manure a year. The distribution of this material among the different animals is about as given in the following table:

TABLE VIII.—MANURE PRODUCED BY FARM ANIMALS

	Number of animals	Number of animal units	Tons fresh manure to an animal	Tons fresh manure	Tons dry matter to an animal	Tons dry manure
Horse	900,000	850,000	10	8,500,000	2.5	2,125,000
Cattle	2,500,000	2,100,000	13	27,300,000	2.2	4,620,000
Sheep	900,000	135,000	6.5	877,000	2.2	297,000
Swine	660,000	100,000	7.5	750,000	2.2	220,000
Poultry ..	10,700,000	107,000	5.	535,000	2.2	235,000
		3,292,000		37,962,000		7,497,000

This production would be sufficient for an application of four and a half tons of fresh manure or nearly one ton of dry matter in manure to each acre of specified crops grown in the State. It is sufficient for an application of one-half that amount for each acre of improved land. This should be a large factor in maintaining the soil. It is worth on the average at least $2.50 a ton on the land and for some crops it is worth much more. It, therefore, has a value greater than the annual budget of the State. But that value is not realized for there is a tremendous loss in handling. It is safe to figure on a loss of one-half the material under the ordinary methods of handling and it may run to two-thirds or three-fourths so that the actual supply of manure to the acre of cropped land is very low, or about one ton a year, or five tons gross or one ton of dry matter, once in five years. While this is a help it is not sufficient, standing alone, to maintain the soil. It is important for the public at large to recognize that while animal husbandry is an aid in maintaining the soil, it is far from adequate even under the best methods of management. For each ton of dry matter in manure, the animal must have consumed about two and a half to three tons of dry matter. If one-half of the dry matter in the manure is lost in handling, for each ton in manure returned to the soil the animal must have consumed from five to six tons.

From these facts it must be apparent how inadequate is animal husbandry to maintain all the soil that is cropped. It should also appear that crop as well as live-stock farming can be made to build up the soil, providing care is taken to return to the soil part of the crop that might be fed to stock. It is just as feasible to maintain the land by straight crop husbandry as when it is combined with animal husbandry. It may be cheaper to grow crops on one acre and haul them to another acre for manurial purposes than to feed low grade or unproductive stock and return only the manure.

These facts give added importance to the use of green crops for manure and to the conservation of the crop residues, the stubble, roots and waste found on every farm. It is not infrequently true that a poorly managed live-stock farm is losing productive capacity faster than a well managed cash crop farm. If animals are kept, they should be of such high producing capacity that they make a profit on the food consumed and the labor and facilities bestowed on them. When one must rely for profit on the manure produced, it is likely to be better to sell cash crops. These facts and the relative price of crop and animal products are some of the things that must be taken into account in a long-range view of the business of agricultural organization, animal husbandry and soil maintenance in any state.

The more varied the products of a state such as New York, the more complicated is the business of farming successfully. The live-stock adjustment in New York is a very complicated and difficult one if the largest present profit and the longest practicable degree of permanence of the productive capacity of the soil are both to be attained.

CATTLE AND DAIRYING (See Figs. 28, 29)

The dairy cow from her numbers and the value of her products commands first place among all farm products in New York. It has already been pointed out that the trend in live-stock has been for many years toward the dairy cow because she is very well suited to utilize the large amount of roughage on the farm and to improve the use of farm labor during the rather long winter period. The confinement to regular hours and continuous daily care involved in dairying has without doubt hampered the attention given to all forms of community activities and doubtless could be traced in their effect on the school, the church and the drift of young people from the farm. There has not been the same state-wide interest on the part of dairymen in gathering in public meetings for their improvement that is manifested among the growers of fruit, for example.

The million and a half dairy cows are mostly concentrated in three main regions: first, in a broad band through the State from north to south reaching from the St. Lawrence Valley; second, around the eastern end of Lake Ontario, and southward between Syracuse and Utica, through the Chenango and Susquehanna Valley regions to the southern boundary; third, in the extreme southwestern three or four counties of the State. The most intense concentration of dairy cattle is between and a little south of Syracuse and Utica. This is the region in which alfalfa is most successfully grown, where the rainfall and snowfall are high and where there are excellent pastures.

From these three main centers of dairying, the number of cows to a square mile decreases very appreciably. Dairying is the dominant industry all through the Mohawk Valley and outside of the fruit-producing district of the Hudson Valley. The general decrease in the intensity of dairying in the Hudson Valley region is due to the small proportion of tillable land and the thin, stony and unproductive character of much of the remaining area. Orange County in the lower part of the State has for many years been a prominent center of milk production. It was into that region that the city of New York first reached for market milk when it had exhausted the territory tributary by short haul. The first special milk train was run from Orange County over the Erie Railroad in 1847.

The keeping of cows is the dominant business along the Canadian line, through the Champlain Valley, and over all the rough hilly sections of the eastern Hudson Valley region. On Long Island, where the population is rapidly increasing, the price of milk is always abnormally high. A fair number of cows has been kept there in the past but is now rapidly decreasing. Dairying is now very largely confined to the southern prong of the eastern end of the island in the region of Freehold and South

Hampton. Most of the island is unsuited to producing forage economically and its geographical situation places a heavy embargo on the importation of concentrated feed because it must pass through the heavily congested traffic region of New York City.

In the centers of most intensive fruit-production, the south shore of Lake Ontario, the grape-belt, the Seneca-Keuka Lakes region and in the Hudson Valley fruit area, the cow has a secondary place. The aggregate area of these regions is relatively small. In the Genesee Valley, dairying is about coördinate with cash crops and with sheep. But even here the cow maintains a place of very considerable importance.

The year-round demand for market milk with the consequent higher price, the opportunity to use the stored roughage of the farm and the possibility of better utilizing labor, have combined to make winter dairying the general practice. The silo has been no small factor in the transition from almost exclusive summer dairying. By winter dairying is meant the practice of arranging for the cows to freshen in fall instead of in the spring. In summer dairying the cows received very little concentrated feed. A very large flow of milk was secured in spring and a fair supply during the summer and fall months, produced by the pasture alone. The cows were wintered almost exclusively on roughage and consequently they gave very little milk in that period. Under this system the cost of producing milk was the lowest possible for the labor and feed employed. By the winter dairying method much of the labor is shifted from summer to winter and a large flow of milk is maintained throughout the winter. This flow is revived when pasture comes on in the spring, so that a much more continuous flow of milk is maintained than under the old "cheese-factory" system.

A few farmers, especially those making the highest grade of milk and cream on tillable land, are inclined to reduce the pasture area and to depend almost exclusively on summer silos and soiling crops. There is no question but that on tillable land a larger amount of nutriment may be secured by this method than by the use of pasture. It should be noticed that this system departs from the chief foundation on which dairying rests, the use by means of the cow of the cheap pasture that is not tillable. It, therefore, entails a higher cost of production.

There is an increasing tendency to supplement the late summer and fall pasture that is frequently short, by the use of silage. The silo has come to be very generally regarded as essential to economical winter dairying and the larger part of commercial dairy farms are now equipped with that structure. In 1917, 40 per cent or 62,000 farms had silos. Corn produces more nutriment to the acre than any other tilled crop grown on the farm, a large part of which is in the stalk, which is difficult to masticate when the fodder is matured and dried. Not only is there loss thereby but the storage of the material in shocks in the field or in bays in the barn entails the probability of further loss. When the crop is preserved in the silo, however, not only is its feeding value nearly all preserved, but it gives the effect of "succulence" in the ration which is so essential to the largest flow of milk.

Another outgrowth of the dairy industry is the equipment of a larger proportion of the dairy farms with commodious and warm barns where the cows are comfortable in winter and where the feeding and milking can be carried on conveniently. Some of these barns, particularly on the farms devoted to the production of certified milk, have been developed to a very high grade of sanitation and labor-saving efficiency. This influence of the dairy toward better buildings is aided by the general custom in the State of storing a large part of the season's crops under cover in the barn, due to the moist summer climate.

The first type of cattle to attain prominence was the Shorthorn, first imported in 1791 to 1796, by Mr. Heaton. In 1817, two bulls, Marquis and Moscow, were brought into the Genesee Valley and were the first animals carrying pedigrees. This strain increased in favor and in 1873 was the occasion of one of the most notable sales of pure blooded cattle ever held. In that year Messrs. Wolcott and Campbell, at New York Mills, near Utica, sold 109 head of cattle that brought an average price of $3,504, and among them was the Eighth Duchess of Geneva, which sold for $40,600, the highest price ever paid for a cow or bull until very recently when as high as $125,000 has been given for animals of the Holstein breed also developed in New York State. As the large eastern cities grew and butter, cheese and milk came into larger demand and the opening up of the Middle West transferred the production of meat animals to that region, the type of cattle in New York changed from the beef to the milk strain. The dominant dairy breed, the Holstein, first brought in by the early Dutch, began to attract popular attention about 1867 to 1875 when there were large importations from Holland. In the earlier days the Devon cattle were introduced into Westchester County and met with much favor. Dutch belted cattle were first imported to America by D. H. Haight, of Goshen, in 1838.

The last available detailed summary of the relative numbers of pure blooded animals of the different breeds was in 1910. At that time, so far as the recorded herds go, 2.9 per cent of all dairy animals were of pure breeding. Undoubtedly, many pure blooded animals were not recorded. In 1918 a rough census of dairy cows showed that approximately 6 per cent were of pure breeding. The total is unquestionably still much below 10 per cent. The total number of pure blooded animals reported in 1910 was 44,423. Of this number, the Holstein formed 72.2 per cent, the Jerseys 14.5, the Guernseys 6.5, and the Ayrshires 4.6, thereby leaving only 2.2 per cent to be distributed among the other breeds. The average number of pure blooded cattle to an owner was 12.1, being largest for the Jersey owners, 18 animals; for the Guernseys owners, it was 12.2 animals and for the Holstein 11.4. On the other hand, observation indicated that the large herds of the State are dominantly made up of the blood of the "black and white" cows.

The commercial size of dairy herds is commonly from twelve to forty cows. There are, of course, many who keep a smaller number to which they look as an important source of income. A few dairies maintain as many as one hundred head of cows and several as high as two hundred head as one unit.

The center for pure breeding of the Holstein is in Madison County. Around this is grouped a half dozen counties of similar prominence in the industry. Solomon Hoxie, a pioneer importer of Holstein, who established the first advance registry record in any breed in 1894, resided at Peterboro in Madison County. This region, with Syracuse as the chief market center, has become widely known among breeders as a source of Holstein stock and many notable auctions of this breed have been held there. There is also a large number of Holstein breeders in the counties of Jefferson and St. Lawrence in the St. Lawrence Valley.

The Jersey breed is not so much concentrated. Several counties, Westchester, Suffolk, Otsego, Delaware and Allegany, have about equal prominence. The Guernsey is best developed in Cayuga, Saratoga and Washington counties. The Ayrshire is coming into prominence in the rougher portions of the State where rustling for pasture is essential. Allegany has the largest number of pure animals of this breed, with St. Lawrence County second. The other three breeds, being more delicate in constitution and habits, are better suited to easier grazing conditions.

The Holstein cow has doubtless attained the present large numbers due to the practice of paying a flat price for milk. She is a large animal and a tremendous milk-producing machine. Holsteins hold most of the world's records both for the total amount of milk and of butter produced in a year, and a large proportion of those records have been made in New York State. The milk, although low in fat, is high in other solids, which is probably the basis of the designation of the breed as the cheese cows. The butter is very mild in flavor and of excellent keeping quality although, like the milk, of an unpopular pale natural color, due to the lack of the rich golden tint that characterizes the product of the Channel breeds. Perhaps another significant fact is the vigor of the animals of this breed, both young and old, and, therefore, the ease with which they may be reared and maintained. The Channel breeds have been relatively more popular than the Holstein with the wealthy man and the cow fancier.

The average production of milk to the cow for all cows in the State is approximately 4500 pounds or a little over 400 gallons. When it is remembered that over 30,000 pounds of milk have been produced by a cow in a year and that many animals in commercial herds yield from 12,000 to 18,000 pounds of milk, it is apparent how low is this average. Further, since abundant figures show that it requires a production of 4500 to 5000 pounds to pay for the cost of keeping a cow for a year, it is evident that many animals are being kept at a loss if a fair charge were made for all items of expense. One-fourth of all cows in the State are kept at a loss; one-half just pay their way and one-fourth return a profit.

The shifts in the numbers of dairy cattle follow the demands of the large urban populations for milk and its products. The milk is used for four main purposes. The proportion required in each group is: raw or market milk and cream for direct human consumption, 40.6 per cent; for the manufacture of butter, 34.2 per cent; for the manufacture of cheese, 20.8 per cent; and for condensed and evaporated milk, 4.4 per cent. These figures are for 1910. The most important change in these proportions is toward an increase in market and condensed milk, and corresponding reductions in the production of butter and cheese. In the ten years preceding 1910, the production of butter decreased 40.0 per cent and that of cheese 18.7 per cent. The heaviest reduction was in the farm-made products. The large urban population in and near the State, especially concentrated in New York City, accounts for this shift in the utilization of milk and the tendency to increase the number of cows. Milk has long been the cheapest animal nutriment and that most economically produced. There is, therefore, a sound basis behind the trend toward cows among farm animals as compared with beef cattle, pork and mutton.

Cattle other than the dairy type are kept in small

numbers. Very few animals are now fed specifically for beef purposes and little attention is given to breeding animals of the beef type. Such feeding of cattle for beef as is practiced is associated with the cash crop systems of farming and, therefore, is best developed in the grain and hay regions of western New York below 1000 feet elevation. Probably the Genesee Valley has much the larger part of the total number. Such animals belong in the region farther west in the United States. Even those few farmers who make a practice of feeding beef animals usually go either to the Buffalo stock-yards or to market centers further west to purchase western range animals in the rough, to be finished on New York pasture and roughage. In the last register of pure-bred stock, 1910, only eighty breeders are recorded who owned a total of 892 pure blooded beef animals. Five breeds are reported in this number, nearly seven-eighths being of the Shorthorn type.

The slow-moving picturesque work oxen so intimately associated with pioneer history and whose numbers were formerly an important item in the list of cattle, disappeared from the census returns in 1900. In 1850, there were reported 178,809 work cattle, and in 1890 only 37,293. Their disappearance from the census returns does not mean that work oxen have entirely disappeared from the State. Draft animals of this type are not uncommonly seen and may be found in nearly every section of the more remote farming districts.

HORSES (See Figs. 28, 29)

The light harness type of horse is the only one in the production of which the State has attained much prominence, and here the record is notable. New York was one of the earliest metropolitan cities of the country and produced the wealth and the leisure in those pre-motor days that were utilized in the fancy coach turnout and the "four-in-hand." It is also expressed in the thoroughbred and finally in the standard bred or trotter which has been so popular in America. The foundation stock of the light harness horse was imported Arabian blood. On the roads on Long Island, some of the earliest races were held. The old Jamaica road or Jericho Turnpike is perhaps the most notable in these annals. The original home of the trotter is Orange County at Goshen where Hambletonian 10, a grandson of the imported Arabian stallion Bellfounder (Messenger) was foaled in 1849 and died in 1876. His monument stands at Chester. He was by far the greatest sire of famous trotting stock the country has produced. Other notable horses have been originated. At Stony Ford in the same county is a monument to Green Mountain Maid, known as the Greatest Mother of Trotters. Membrino Chief was foaled across the Hudson River in Dutchess County in 1844, and Ethan Allen, a notable sire in the Morgan strain and son of Justin Morgan, was also foaled in the Hudson Valley. Goshen, in Orange County, was known as the "Lexington of the North." The Puritan attitude toward horse racing is reflected in the fact that as early as 1802 horse racing and trotting was prohibited by law. In 1821 this law was amended to permit trotting and running races in Queens County.

Most of New York has a rough topography, the roads as well as the fields often being steep. Light active animals of endurance are considered by many farmers to succeed better in general farm use than do heavy horses. The heavy horse, adapted to deep pastures, rich feed, and slow movement, is better suited to relatively level than to broken hilly country. The earlier glamor of the speed horse led to the extensive use of sires of the roadster rather than the draft type, on the ordinary light mares of the rural districts. Consequently, the prevailing type of horse in New York is a rather light animal showing much of the blood of the speedway, of the coach and the chase. It has been estimated that 75 per cent of the horses in the country are of a size between 1000 and 1400 pounds. They are a medium sized dual-purpose animal that can do a fair day's work on the farm, and that can keep up a fair gait over the road with a small load or attached to a light wagon.

The New York farm, perhaps largely because of its hills, and also its history, is still in the era of small fields and small implements. The single bottom plow and the light binder and mower predominate rather than the big broad-cutting tools. Whether the automobile and the improved roads will effect a larger trend toward the heavier draft type of horses remains to be seen. It seems likely that there will be an increase in the proportion of the heavier horse that is preëminently adapted to heavy draft purposes. It is clear that the heavier horses are increasing in favor. But in those parts of the State composed of steep fields and roads, a fairly light type of horse is likely to continue in favor. On the other hand, the farm tractor of which there were 3000 in 1917, and rapidly increasing, is likely further to cut into the demand for heavy horses.

New York is a horse-consuming rather than a horse-producing state. In 1910 there were on farms in the State 591,008 horses of which 562,310 were mature animals. There were approximately 303,000 horses not on farms, that is, in use in towns and cities. This makes a rough total of about 900,000 horses. The number of horses on farms decreased in each of the last two ten-year periods from a maximum number in 1890, when there were 713,384. It is estimated that the average horse in the city is good for five to six years, while the average horse in the country will give twelve to thirteen years of service. These figures mean an annual consumption of about 60,000 horses in each group or a total of 120,000 that are used up each year.

The annual production of horses in New York in 1910 was only 25,083 animals, or less than half enough to supply the rural needs. Dealers have estimated that as high as 95,000 horses are imported annually. The business of dealing in horses is, therefore, large and some very large concerns have been

Plate VI. Representative fruit plantations in New York. Young orchard and vineyard above; bearing apples in bloom below.

developed in New York and Buffalo, which are the chief centers of exchange. It is stated that one concern in New York City has handled from 35,000 to 40,000 horses annually. That the production of horses has decreased is indicated by the fact that in 1890, the year of the largest number of horses, the production was nearly twice that of 1910, while the total number of horses on farms was only 17 per cent greater.

The distribution of horses follows closely the improved land and averages fifteen acres of land in specified crops for each mature horse. The proportion varies with the intensity of the farm practices. In all those regions primarily devoted to dairying, there is one horse for each twenty-six to thirty-five acres of improved land. In the counties near large cities, it is one horse for eighteen to twenty-five acres and in Nassau County, largely devoted to market-gardening, there is one horse for each 11.3 acres of improved land, or 5.8 acres of land in specified crops.

The increase in the number of horses on farms since the census of 1910 is doubtless traceable to the tendency to increase the intensity of farm operations.

The draft breed of horses in most favor is the Percheron. His clean limbs, intelligent and alert, active movements in many ways adapt him to conditions as they exist in the better farming sections of the State. In 1910, an incomplete register of pure-blood stock showed forty-eight breeders of draft horses. They were distributed among 281 owners. The largest number of breeders was in Allegany and

Steuben counties. The largest single lot of pure-blood animals was in Ontario County. Syracuse has for many years been a center to which Percheron horses were imported from France.

Next to the Percheron in prominence is the Belgian horse. The Clydesdale is increasing in favor because of his long pastern and his consequent better performance and durability on the road. All the other heavy breeds have some fanciers.

Among light horses, the persistence of interest in speed and fancy-driving is shown by the fact that out of 198 breeders who owned 1505 pure-blood animals in 1910, there were 21 owners and 284 horses of the thoroughbred class, 28 owners and 188 animals of the American Trotter, and 13 owners and 65 animals of the Morgan strain.

Among coach horses, the Hackney has been the most popular breed, with the French Coach and the German Coach in a poor second and third position as regards pure breeding on the farm. However, in the metropolitan cities, the Coach horses, the Saddle Horse and the Hunter have claimed most attention. The great horse shows that were held annually in New York for many years in Madison Square Garden were the Mecca to which horse fanciers the country over made annual pilgrimage and these shows have also been social events of no small proportions. While the light harness classes have claimed most attention before audiences free to indulge their whims, the heavy harness horses and heavy draft turn-outs have long received the attention of concerns in the city using many heavy horses for trucking and other heavy hauling. For these reasons also, the shows were of interest to highly specialized breeders and importers in all lines of the horse industry. With the advent of the automobile and the annual automobile show in New York, the attention given to horse shows has been much displaced. As a result of the large introduction of motor cars, it appears from general observation that there has been a very considerable decline in interest in horses of all classes since the last census of pure-blood animals was taken. The popular opinion that the automobile truck and tractor will replace the horse, however, is erroneous. Certain work will be performed by the gas engine, relieving the horse from these particular lines; but the increase of acreage and yield following the employment of mechanical power will make more work of other kinds for horses. There is no way yet foreseen whereby the farm horse can be displaced. He may be taken off the road, but he is still the most available power for much of the work of agriculture.

MULES (See Figs. 28, 29)

Mules have never been very popular in New York. In 1880 there were a thousand more individuals on farms than in 1910. In the last federal census year, there were 3,490 mules on farms and 4,052 in other employment, both of which figures represent an increase from the preceding census. The increase on farms, 88 per cent, was four times that in cities. Farmers who have had experience with mules are very partial to them as they are safer than horses in the hands of poor horsemen who are common among present day farm laborers. The production of mules is small. This is indicated by the fact that there were only 192 yearlings on farms in 1910. These figures should be noted in connection with the number of asses and burrows kept in the State. This was 284, probably nearly all of the former type, the number of burrows being almost negligible.

SHEEP (See Figs. 30, 31)

There has been a nearly continuous and regular decrease in the number of sheep on the farms of the State since 1840. Only in the census year of 1900 was there an increase in number over the preceding ten-year period, probably in response to the general demand for wool and the low cost of feed. The following table gives the number of sheep kept and the amount of wool produced by periods since 1840. About two-thirds of these are mature animals, and the remainder are lambs.

TABLE IX. NUMBER OF SHEEP AND AMOUNT OF WOOL PRODUCED SINCE 1840

	Number of sheep	Wool pounds		Number of sheep	Wool pounds
1840	5,118,777				
1850	3,453,241	10,071,000	1890	1,528,979	6,715,000
1860	2,617,855	9,454,000	1900	1,747,062	6,674,000
1870	2,181,587	10,599,000	1910	933,775	4,235,000
1880	1,715,180	8,827,000	1917	587,132	

The production of wool to an animal, estimating two-thirds of the total number as mature animals

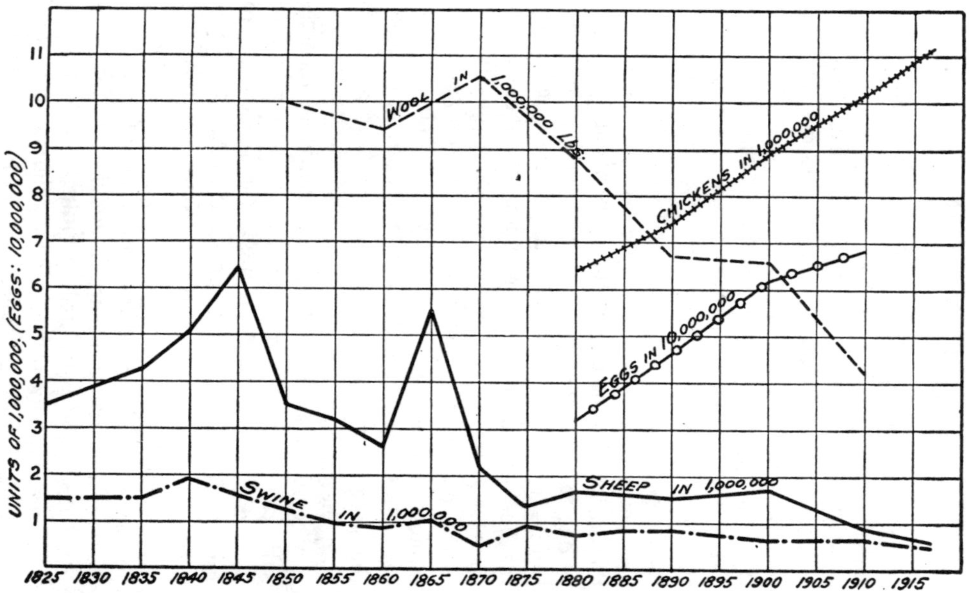

Fig. 30. Graphs showing by five-year periods the numbers of swine, sheep, and chickens and the production of wool and eggs.

producing a fleece, varied considerably during that period. In 1850, it was 4.4 pounds. It increased regularly to 7.7 pounds in 1880 and then decreased to 5.7 pounds in 1900, but in 1910 it was only 6.8 pounds an animal on the same basis of calculation. Based on the actual number of fleeces reported which was 573,611, the average weight was 7.4 pounds in 1910 and 6.4 in 1900. The most notable point about these figures is the large increase in average size of fleece over that in 1850. It has been suggested that the decrease in the size of the fleece about 1900 may be due to a considerable change in the type of sheep kept. In the earlier years, they were dominantly the fine wool breeds having much oil in the fleece. In later years the coarser wooled sheep prevail.

The dominant cause of the heavy decline in the number of sheep has undoubtedly been their large increase in the middle states and on the western ranges, the so-called short grass region. Another cause has been the development of the sheep industry in such new countries as Australia and Argentina. As a result, the price of mutton and wool was very low. A mature sheep sold in 1893 for as low as 75 cents. The price of mutton in 1895 was 3.1 cents a pound. Dairying has paid in New York relatively better than sheep, due to the excellent local market for milk. Added to this combination of conditions has been the serious menace from dogs that for many years have been permitted to multiply and run free with scarcely any restraint or regulation and which have cost the sheep industry millions of dollars in actual animals destroyed and injured. Figures collected from county and town records in 1913 indicated that at least 1500 sheep had been killed by dogs and paid for by the towns in that year. It is more accurately estimated at six times that number of animals, counting the animals unreported and the ultimate loss in vigor, from attacks, so that the true figures should be 9000 sheep, not to mention the far

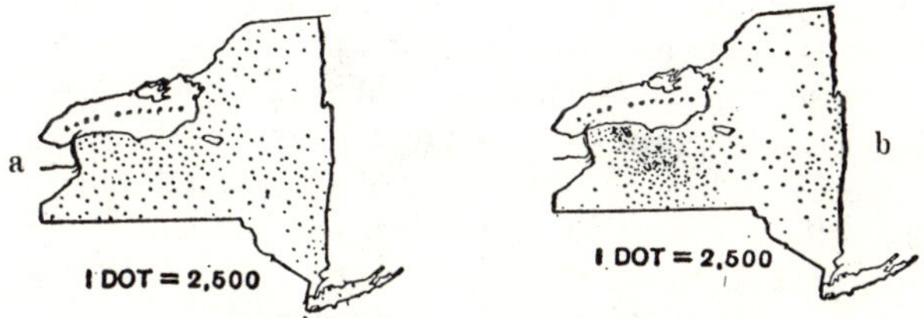

Fig. 31. Maps showing a. number and distribution of sheep; b. swine in 1910.

greater damage to the flock from fatigue and fright. Provision is now made in the State law for compensation covering this latter type of damage to the flock. The larger demand for mutton and wool, and the failure of competing regions to develop in proportion to that demand, have led to an increase in the price both of mutton and wool that is rapidly changing conditions in favor of sheep-raising in New York, wherever the cow does not fit the system of farm management or the markets.

So far as farm conditions alone are concerned, New York has much land that is especially suited to the production of sheep. They are adapted to the rough and more or less waste land over which they may graze and collect the low-grade herbage. They are useful to clean up weeds and much brush. They will also utilize much by-product material not so well adapted to consumption by cattle. They require less shelter than dairy cows. Many farms in the more remote districts near the border line of abandonment will provide ample buildings suitable for sheep-raising with very little change. Elimination of fences and the opening up of larger ranges in those regions are both desirable and practicable. When these grazing areas are combined with those of better soil on the level part of the upland and in the valleys to provide winter feed, an excellent situation is afforded for the development of the sheep industry with a minimum requirement of labor.

Sheep and cows are not raised in large numbers in the same region. Sheep are seldom found on commercial dairy farms. The two main regions of sheep-raising are: the Genesee Valley and Ontario Lake Shore section, the bean-growing area; and the eastern Hudson and the Champlain Valley districts. The Hudson Valley is somewhat extended westward into the lower Mohawk Valley.

In the Genesee Valley region, in addition to the general roughage and pasture, the bean straw is especially well adapted for consumption by sheep. Sheep also fit better than cows into the organization of the fruit and cash crop farm. Nearly 40 per cent of all sheep in the State in 1910 were in the eight counties comprised in this district with the main valley of the Genesee River as the center of production. The former large development of the industry of sheep-raising in New York was not confined to these districts. Sheep were kept in large numbers in every county. One may surmise that this wide distribution had some relation to the fact that the spinning and weaving of woolen garments were then a home art and that the wool came from the farm flock.

The introduction of pure-breed sheep began about the same time as that of Shorthorn cattle, 1790–1795. The first Merino sheep were two ewes and a ram imported to Boston by William Foster in 1793 and by him presented to a friend in the Pawling Valley, Dutchess County. Not knowing their value, they were killed for meat. Robert Livingston introduced four head of pure blood Merino sheep into the Hudson Valley in 1801 that had a very large influence on the sheep in that locality. The Southdown was brought into Fayette, Lewis County, by Dr. Rose in 1803. Other first American introductions were the Cotswold, by C. Dunn near Albany in 1823, and the Cheviots at Delhi by Robert Young in 1838. In the first part of that century, New York offered a premium of $50 to any person bringing a Merino ram into a county. About the same time the State agreed to loan $5000 to any person who would establish a woolen factory, and a premium of $150 was offered for the best two yards of woolen cloth. In 1828, tariff protection on wool was demanded.

An incomplete register of sheep in New York

records fifteen breeds with two or more owners of each breed. Out of 420 breeders of pure blooded sheep who owned 13,893 animals in 1910, the Shropshire leads with 137 owners and 5725 animals. The State is regarded as peculiarly suited to the Shropshire. The three Downs breeds were represented by 111 owners and 2876 animals, of which 1421 were Hampshires, 709 Oxfords, and 746 Southdowns. The Merinos stand next in importance with 35 owners and 2470 animals, while the Dorsets had 41 owners but only 1034 animals.

Such authorities on sheep husbandry as the late Joseph E. Wing recommend for New York the keeping of sheep, by which he means the production of wool rather than of mutton. At the same time, other authorities have emphasized the market facilities of the State for mutton and especially for lamb. The commodious warm barns available on many farms, together with the need for animals to consume roughage and to provide winter employment where dairying is not practiced, favor a moderate development of the industry of winter or hot-house lamb production. H. A. Hopper says that "the maximum returns under eastern conditions will be secured from sheep giving as heavy a fleece of good staple wool as is consistent with the economical production of the mutton type demanded by the market. In other words, the dual type of sheep that is able to produce a fair sized fleece and at the same time develop a good quality of meat at a fair cost, is the one best suited to conditions in New York."

Flocks of grade animals rather than pure bloods seem best for the average farmer. They need to be hardy and good rustlers. For this purpose, a foundation of Merino crossed on Down or Dorset breed is suggested by those experienced in sheep husbandry.

GOATS

Goats, the number of which was 3475 in 1910, are widely and rather uniformly distributed throughout the State. There are a few milch goats. For several years a herd of fifteen was kept by the State Experiment Station at Geneva and a study made of the cost of production of goat's milk and of its food properties. The average yearly production of ten animals during three years was 800 pounds of milk. The range was from 307 to 1845 pounds. The average food cost of the milk was four cents a quart which should be compared with 0.92 cent a quart in a herd of twenty-five Jerseys during the same period. This was in the years 1910–1912 inclusive. The total solids during the summer months was about 11.5 per cent. The percentage of fat was about 3.6, but in both respects a large variation was noticed for individual animals. No appreciable difference from cow's milk in composition was found. As a food for infants it has one advantage due to the finer curd it forms which makes it more easily digested than cow's milk. On this showing of efficiency there is no promise that goats will be a substitute for cows except in the few circumstances where the latter can not be kept.

SWINE (See Figs. 30, 31)

There has been much less fluctuation in the number of swine kept in the State than in sheep. In 1850 there was one swine for every 3.5 sheep; in 1910 there was one for each 1.4 sheep. There has been a marked decline in the number of swine as shown by the following table, which gives the numbers by census periods from 1850 forward:

TABLE X. NUMBER OF SWINE FROM 1850 TO 1917

Year	Number	Year	Number
1850	1,018,252	1890	843,342
1860	910,178	1900	676,639
1870	518,251	1910	666,179
1880	751,907	1917	435,908 *

* Fifty-six counties only reported.

Of the total number in 1910, a little more than half, or 55 per cent, were mature animals, while for sheep the proportion of mature animals was 65 per cent.

Hogs in New York are preëminently the scavenger animals. As has been pointed out, they make gain in weight more efficiently than any other type of livestock. The average distribution is one animal for each three rural inhabitants, and ranged from one for each 1.2 persons in Nassau and 1.4 persons in Cayuga county to one for each 3.5 persons in Madison and one for each 3.6 persons in Chenango County.

Hogs are not kept in New York to consume crops that could be sold for cash as is the custom in the corn-belt. In Illinois, there is an average of eighteen swine to a farm, and in Iowa thirty-five. In New York the average is less than two to a farm. It is interesting to note that swine are least numerous in the leading dairy sections, especially the market milk regions.

It is evident that in New York hogs are kept primarily to provide pork for the home larder. Their nearest competitor as a scavenger has long been poultry. Up to the census of 1910, the actual intensity of swine-raising was greater than that of poultry. About that year, the number of poultry became equal to or slightly larger than that of swine. The intensity was then approximately one-half animal unit to a farm. With the average number of swine at two to a farm, the average number of animals one year old or over is only about one. There is much less than one brood sow to a farm and consequently the litter of each sow is likely to be a distribution among several farms.

The region of largest swine-production is along the south shore of the Great Lakes and eastward through the Mohawk Valley. There seems to be no reason to expect any large increase in the number of swine. They are urged by some fanciers for production on green forage and pasture. But other animals, such as the dairy cow, can use this material more efficiently. Warren has shown by calculations from the relative price of crops and of pork that there is an effective price tariff in favor of hog-production in the corn-growing states amounting to one-third to one-fourth of the market value of the animal.

Breeds of swine, unlike sneep, have been largely developed in America. Three breeds have been originated here, one of which has taken a leading place. The Duroc-Jersey was developed by Colonel F. D. Curtis, of Kirby Homestead, in Saratoga County, about 1823, and he also originated the Curtis strain of the Victoria breed in 1850. The Cheshire pig was developed in Jefferson County in 1835. The leading breed in the State has generally been the Berkshire, an English stock of which there was an important introduction near Albany by Sidney Howes in 1832.

Thirteen breeds of swine are represented in the State. Of these, seven have considerable numbers. However, the Berkshires distance all others in popularity. Out of a partial register in 1910 of 431 breeders and 5102 animals, 225 breeders and 2903 animals were of the Berkshire breed. The Chester Whites, Cheshires, Yorkshires, Poland-Chinas, and Duroc-Jerseys ranked in numbers in the order named. The activity, vigor, prolificacy, early maturity and fine quality of the meat of the Berkshires admirably adapt them to the average farm of the State.

POULTRY (See Figs. 30, 32)

The hen vies with the pig as the preferred farm gleaner and scavenger. In the census of 1910, for the first time, the animal unit strength of hens exceeded that of hogs. In this calculation, one hundred chickens and five mature hogs are rated as one animal unit. Each animal is found on the average farm to the extent of about one-half animal unit, the pig now falling considerably under that figure. In former years the pig has run up to a strength of six-tenths of a unit to a farm and in that same year,

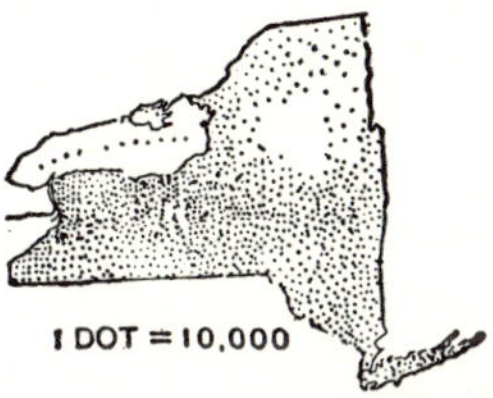

Fig. 32. Number and distribution of poultry in 1910.

1890, the hen had a strength of only three-tenths of a unit to a farm.

The number of poultry has been increasing regularly and rather rapidly. The figures for the last three census periods are given in the following table:

TABLE XI. NUMBER OF POULTRY, 1880 TO 1917

Year	Number	Year	Number
1880	6,448,886	1910	10,265,934
1890	8,421,667	1917	11,252,000
1900	8,964,736		

The rank of the State in poultry production for meat, is not high, being tenth in 1910. In egg production, however, the rank was eighth. This is well shown by the number of eggs produced by a hen: in New York it was seven dozen, in Illinois 4.7 dozen, Iowa 5.3, in Missouri 5.8 dozen and in Ohio 5.5.

These are the leading poultry-producing states. Iowa, the leader, and Illinois in second place each have more than twice as many chickens as New York and a larger number to a farm. The 50 to a farm in New York must be compared with 45 in the United States as a whole, 108 in Iowa, 81 in Illinois, 75 in Missouri, and 64 in Ohio. Poultry, like pigs, follow cheap feed in their main distribution; therefore, the largest numbers have been in the corn-growing region. It is evident that the farm flock is the big factor in the poultry business in New York. Every farm has a few hens just as nearly every farm has a pig or two. The hen is a side line on the farm usually left to be cared for by the women and children who frequently derive their "pin money" from a little extra attention to the flock. The number of small commercial plants is multiplying and it is a favorite business of the "back-to-the-lander" and the "commuter." This is illustrated by the distribution of chickens in the State. In general, like that of pigs, it is in about the same proportion as the rural population. They are most numerous along the Great Lakes and through the Mohawk and Hudson valleys, thinning out in the more sparsely settled sections of hilly or rough country. However, there is a disproportionate increase in the number of chickens in the southeastern part of the State. It is the only type of food-producing live-stock in considerable numbers on Long Island. Many persons have come out from the cities and started a small poultry business. It has been said that there are more failures in poultry production than in any other line of agriculture, probably due to the number of novices that have engaged in the business. It is also a complicated business, many persons who have successfully kept a few hens having failed in larger enterprises of the same sort. A few hens kept on the farm as a side line very well care for themselves with little attention. Natural forces protect them and provide their feed. In larger flocks, however, this is not the case. On the other hand, many farmers who have begun to record the cost of keeping different farm animals and have counted the returns, have found that the hens were the only type of stock that were paying.

New York City in particular, and in fact all the cities in the region to a large extent, demand white eggs. They pay a premium for "chalk white" eggs, that ranges from two or three cents in April to twelve or fifteen cents in November over brown eggs. Boston is the market for brown eggs. Consequently, hens that lay white eggs are far the most common in New York State. This means the Mediterranean breeds with the White Leghorns far in the lead. This is particularly true wherever attention is given to the commercial aspects of poultry-keeping. Available data indicate that over two-thirds of all birds are of the White Leghorn strain. On the other hand, on many farms where both eggs and meat for home use are desired, one of the heavier breeds that lay brown eggs is kept.

Much attention is given in the schools of the State both of secondary and college grade to improving farm poultry, and especially its egg-laying capacity. At the State College of Agriculture at Ithaca, a White Leghorn hen has laid as many as 327 eggs in a year and as many as 1229 in a lifetime of six years. These records are not equal to the best that have been made in other states where the record is 329 in one year.

It is evident that there is still much opportunity for improvement by breeding, selection and care. The latest development in the poultry business is the introduction of artificial light for a period in the morning and evening which result quite uniformly in a larger production of eggs, especially in the winter months.

Turkeys, ducks, geese and guinea fowls are of minor importance. A farm in Niagara County devoted to duck-raising produces more than 50,000 mature birds in a year and there are several large duck farms in the lower southeastern part of the State and especially on eastern Long Island. In the vicinity of Eastport, single ranches fatten as many as 100,000 ducks in a year.

Turkeys, once a common part of the poultry on every farm, are now very scarce. They are found only on the farms in the more remote districts where they can have a free range.

In 1910 the total number of turkeys, ducks and geese was only 300,755 and the State ranked fifteenth. This was much less than half the number in the leading states. Being essentially meat birds and requiring even more special attention than chickens, they have been dropping away in numbers in New York and have followed the meat end of the chicken business into the middle western states.

BEES AND HONEY

Wheeler Dennison Wright in a bulletin of the State Department of Agriculture published in 1913 says: "Beekeeping on a commercial scale is far from being a royal road to wealth as pictured by some, but requires as much labor, diligence and attention to details as many other lines of business. . . . In a small way, beekeeping is quite popular as a recreation for persons of sedentary habits, also as a light employment for invalids. The Empire State ranks high in the production of surplus honey, and numbers its beekeeping specialists by the hundreds. Its honey yielding flora is extensive and diversified."

A considerable number of beekeepers have colonies of forty to one hundred stands scattered about the region of their operations at intervals of three to four miles or more. According to the census of 1910, there were 156,360 colonies in the State. The rank of New York was sixth, having advanced from eighth place in 1900. In spite of this advance in rank, there were not as many colonies in the State by 30,000 as in 1900, so that other states must have experienced a heavy decline in number.

The production of honey was 3,191,733 pounds and of wax 43,198. In the former the State held second rank, being exceeded only by California which is reported to produce three times the amount made in

New York. Apparently, the general standard of honey production in New York as well as in the United States at large, is low as compared with California where the yield of honey was 50 pounds to a colony a year, while it was 20 pounds in New York and 16 pounds as the average for the country. A woman beekeeper in the State who had 170 colonies reports the production of 11,500 pounds of comb honey and 2,200 pounds of extracted honey in one season. This is at the rate of 80 pounds a colony.

Bees are very widely and quite evenly distributed through the populated sections of the State. Like pigs and poultry, they follow the distribution of rural population. They are the supreme gleaners, retrieving human nutriment in an amount and by industry that cannot fail to be the marvel of the average person. Contrary to the former two types of animals, there appears to be a greater concentration in the more remote farming districts rather than in the suburban sections.

Bees are fostered by State and County beekeepers' associations and have the protection of the State laws. One is not permitted under the State law to keep colonies infected with the disease known as foul- or black-brood. Further, fruit-growers are prohibited from spraying with poisonous or otherwise injurious sprays when the fruit is in full bloom because of the injury done to bees.

CHAPTER VII

MARKETS AND MARKETING

THE disposition of a commodity, whether it be apples or automobiles, manual labor or mental product, depends on two fundamental elements: there must be a place of need, and a means of putting the commodity in touch with the need. These underlying facts are not to be confused with the large mass of details that may affect the character or the volume of the need, or with the means for bringing the supply of a particular thing to the place where it is to be used.

In the discussion of marketing problems in New York, it will be well to keep in mind these general distinctions as a means of separation between the underlying facts and the numerous details of greater or less importance that may need to be adjusted to secure satisfactory market conditions. The underlying factors are natural and are not much affected by individual effort. The details of use, of grade, of market customs, of methods of transport and handling may readily be changed and adjusted. The fundamental facts are the nature of the products and the manner of their use.

The commodities of agriculture are things to eat, to wear and to use, mostly in an elemental way.

Therefore, they touch the whole people and in a study of market facilities, account may first be taken of the adjacent population to be served. In this respect, New York farmers stand in a peculiar, in fact in a specially advantageous, position. They are close to the largest population of any similar area on the western hemisphere.

The states and provinces touching New York and including its own area have an aggregate population of thirty million. A circle described around Syracuse as a center and with a radius of five hundred miles would include not only this population but would add most of that of the states of Ohio, West Virginia, Maryland, Virginia, Delaware, Rhode Island, New Hampshire and Maine, with a total population of about fifty million or nearly one-half of that on the North American continent. Syracuse has been aptly termed the hub city, and New York is truly a hub State to the population and industry of the northeast country. Of this population, the largest proportion of any division of the continent is engaged in other than agricultural pursuits. The region is essentially an urban one. It also has the largest proportion of the population engaged in manufacturing of any division. Adjacent on the south is the great mining region of Pennsylvania. Within this eastern district there are twenty cities with a population of one hundred thousand or more. All these facts go to make up the consuming capacity of that division of the country for the things of the farm. Of this population, approximately 70 per cent is

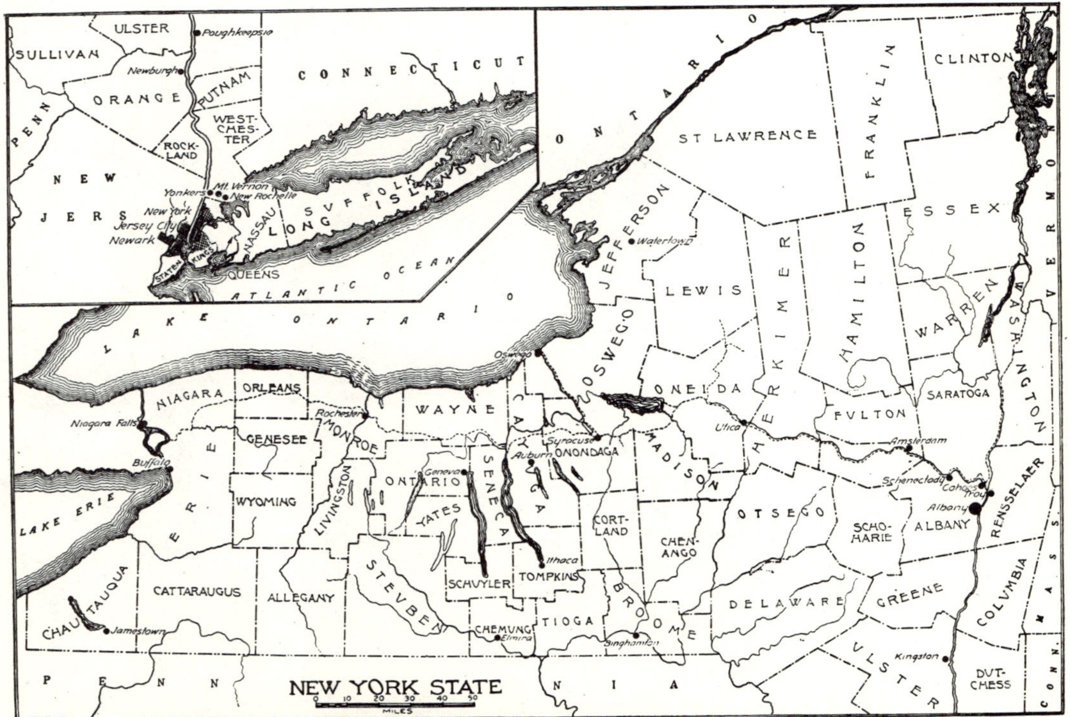

Fig. 33. General map of New York State showing drainage systems, counties, and chief cities.

strictly urban according to the method of distinction of the census, and to that proportion may be added fairly a further 10 per cent to cover persons living in towns and village under two thousand five hundred population. This reveals about 80 per cent or four-fifths of the population as urban and therefore engaged in other than agricultural pursuits and entirely dependent on the products of the farms.

The population of New York State as a whole is wealthy. The region is the center of wealth of the continent. Added to this is the tremendous transient population of business and pleasure, and of whose presence the immense hotel capacity of the larger cities is evidence. These are the big facts with reference to the consumption capacity of the population within convenient reach of the farmer located in New York, and this proximity should constitute a natural subsidy to him in marketing farm produce.

In addition to the population that may be said to be at the door of the New York farmer, there is the further large mass that may more readily be reached by him than by the farmers of any other part of the country,— the foreign centers of population reached directly by ships that clear from the large maritime cities within this five hundred mile zone. All Europe is at his door as a result of modern means of transportation and the preservation of perishable farm products.

New York, the chief American port and one of the three largest shipping ports in the world, is a little over thirty-five hundred miles from Liverpool and London. Added to this are the ports of Boston and Philadelphia that are among the first five in the country. In 1910 the freight tonnage that cleared from New York for foreign ports was approximately 652,000,000 and the imports amounted to 936,000,000 tons. From one-third to one-half of all the foreign trade of the United States passes through the port of New York. The cost of freight a gross ton of staple commodities from New York to Liverpool before the war was from two to four dollars, which was equivalent to the average railway freight for a distance of three to four hundred miles, say from New York to Buffalo. There are, of course, many differences in the make-up of transportation charges by ocean freight and by rail, but these figures will serve as a general basis of comparison and indicate the way in which local, in the sense of home, markets are expanded by facilities for foreign shipment.

The consuming capacity of this territory is immense and it only remains properly to direct it and provide the means for efficiently and economically reaching the market. This involves among other things the means of transportation and communication.

TRANSPORTATION AND MARKETS

The transportation facilities may be divided into three groups: highways, railroads, and waterways. The first determines the access to the local shipping point. The latter two are the means of reaching the more remote centers. These different agencies co-operate with each other in facilitating transportation.

New York State has 8550 miles of steam railroad and about 5000 miles of electric railroad, a large part of which, of course, is in cities. The proportion of this latter that may be assigned to rural districts may be placed roughly at one-fifth or 1000 miles. Assuming that 550 miles of the steam road is in cities, the net mileage of railroads in the country districts of the State is 8000 or about one for each five square miles of area or one mile for each 1000 acres of specified crops.

Unfortunately, this mileage is not uniformly distributed and in several places there are double and triple lines approximately parallel and near each other, which correspondingly reduces the railway service to other communities. The greatest thoroughfare of rail traffic in the country is the course up the Hudson Valley and west through the Mohawk Valley, over the Ontario plain to Buffalo and thence to Cleveland and the west. Another important line of travel lies across the southern part of the State leading up from New York through the Delaware and Susquehanna River valleys and thence up the Chemung Valley and over the divide into the Allegheny Valley on a course to the Middle West.

The hilly topography, together with the peculiar deep through-valleys that have been developed in all the more elevated parts of the State, have been responsible for this concentration of the main lines of transportation in the valleys. The cross valleys that interlace with these main ones are also occupied by railroads in many cases. A network of lines spreads over the productive parts of the State in a fairly uniform system within the limits of these valleys.

A consequence of this general arrangement of the railroads in the valleys is the remoteness of those areas that lie on the hills whose practical distance from shipping stations is frequently made much greater by the steepness of the grades that must be overcome in reaching the station.

The course the railroads are obliged to follow also influences the routes of travel and transport and may make them roundabout. This interferes with the free exchange of shipments between railroads to secure direct transit. The deflection of lines of travel due to topography is not as large as in Pennsylvania but it is far larger than in states like Ohio or Illinois. All this is reflected in the tariffs which, until recently readjusted under federal supervision, were widely different in various parts of the State.

Examination of any good map of New York will show the location of the railroads. The electric roads, which are a recent development, are being pieced together to form important through systems of transportation. With the exception of a short gap of thirty miles, it is now possible to travel by electric road from the western boundary of the State over the Erie plain to Buffalo, then to Rochester by way of a succession of important towns on the edge of the Ontario plain, thence to Syracuse, Utica and Albany.

The gap occurs from Little Falls to Fonda in the Mohawk Valley. At Albany one may travel northward to Glens Falls by two routes, and southward as far as Hudson. Several important spurs of electric road lead off north and south from this general line, for example at Syracuse to Oswego and Auburn, and at Herkimer to Oneonta. All this line is broken into a series of pieces by separate ownership without joint traffic arrangements, so that it is serviceable only for local rather than long distance shipment. It is used primarily to get from the rural districts to the nearest large cities. Hence the name inter-urban. This electric service is frequently affiliated with city lines or with steam roads.

In the southern part of the State, there are several pieces of inter-urban electric railroad. The more extensive of these center at Goshen, Elmira, Olean and Jamestown.

Inland water transportation is provided on the Hudson River as far north as Albany and a large amount of local produce is shipped by this means to New York. The Barge Canal system, represented by the Erie Canal with 339 miles, the Champlain Canal 61 miles, the Oswego Canal 23 miles and the Cayuga-Seneca Canal 23 miles, with a total length of 446 miles, is in such a state of incompleteness that its service as a means of shipping agricultural produce can only be guessed. In its length will be some of the largest canal structures in the world including the highest single lift lock, namely 40½ feet, at Little

Falls. The most notable series of locks is at Waterford near Troy where the five locks have a combined lift of 169 feet. Another unique feature is the movable dams to regulate the level of water in the Mohawk River. The total cost of this system of canals is $150,000,000, borne by the State of New York. If the Hudson River below Troy is included with the canal system, a strip of country within two miles of the waterway on either side includes 73½ per cent of the population of the State and within twenty miles is 87 per cent of the population. Naturally, the slower transit makes the canal suited to only the less perishable staple crops that may move slowly to market. The hundred and fifty million dollars that will ultimately be expended on these canals and their terminals to enlarge them to power barge capacity with a depth of twelve feet and a width of seventy-five feet, will make a system of waterways that will be especially serviceable in the transfer of heavy products from the West through the Great Lakes. But its value for the movement of farm produce is problematic. The big disadvantage of transportation through the Great Lakes and the St. Lawrence River by way of the Welland Canal between Erie and Ontario lakes is the fact that the upper course of the river is closed by ice from October to May.

The rates and operation of railroads within the State come under the supervision of the Public Service Commission, which exercises a potent influence to secure reasonable service. By investigation of delays and improper practices, this Commission has adjusted many complaints of shippers of agricultural produce, although, of course, it cannot make regulations or provide facilities that are not reasonably warranted by the freight to be handled.

Less than a quarter of the products of the farm are ever carried by railroad or other long-distance method of transport. They are hauled to the adjacent towns and cities and are used locally. They are hauled on the highways. The first movement from the farm is over the country roads and the cost of transportation from the farm to the station or to the local point of consumption may easily be as much as the freight cost to a remote market because of the small load that can be carried. The character of the rural highways plays a large part in the marketing of produce, and also in returning supplies to the farm.

The average haul of farm products to market by wagon was found by a special investigation by the United States Department of Agriculture to be seven and three-tenths miles. The average cost a ton to get to market for eleven important crops was one dollar and ninety cents. The cost a ton for a mile of transportation was twenty-six cents. The cost of railroad freight a ton mile was at the same time from one to two cents. Both figures would be much higher at this date. These facts emphasize the importance of rural highways in the problem of getting produce to market. The length of haul, the cost a ton and the cost to the ton-mile in New York as determined by the United States Department of Agriculture in 1907 were as follows:

TABLE XII. COST OF HAULING FARM PRODUCTS TO MARKET

Commodity	Average distance to market	Average cost per ton	Average cost per ton-mile	Value per ton
Apples	7.4	$1.80	24.3	$34
Beans	3.9	1.60	41.	66
Buckwheat	9.2	2.20	24.	40
Corn	8.6	2.40	28.	25
Fruits	6.0	1.60	27.	75
Hay	6.6	1.60	24.	15
Hops	10.6	2.80	26.	240
Oats	7.8	1.60	20.	25
Potatoes	7.1	1.80	25.	20
Vegetables	7.0	1.40	20.	30
Wheat	5.9	1.80	30.	33

From these figures it is evident that the cost of moving farm produce on the average country road in New York is about twelve to fifteen times as great as on railroad.

The mileage of public roads in the State is 80,000 or one and six-tenths miles to a square mile of total area and one mile for each 100 acres of specified crops. About a third of the State can be set aside as rough wooded country with very poor road service and needing very little, so that the mileage of road a square mile of improved area is two and a half miles. Of the total mileage, 50,000 or 62 per cent is rated as improved and about 17,000 miles is macadamized.

In 1898, New York adopted a system of highway improvement under an arrangement by which the

State pays 50 per cent of the cost of such improvement, the county 35 per cent and the town or village the remainder. The roads that receive this improvement are laid out in state, county and town systems, the first being designed as trunk highways for through travel, the second division constitute the main local highways, while the third comprises the local·highways of secondary importance. The State has authorized a bond issue of one hundred million dollars to pay its portion of the cost of this improvement and the expenditure of the second half of the appropriation is well progressed. The construction is accomplished in sections from year to year by designation of the local authorities, the county boards of supervisors, in conference with the state highway department, the apportionment being divided in proportion to the total mileage of highway. A resident highway engineer is maintained in each county to supervise and inspect the construction which is done under contract. The local maintenance of country roads is under the money rather than the old service system which latter method has proved very inefficient.

The automobile has been a large factor in promoting the improvement of highways in New York, as in all other parts of the country. It is having a large influence on the rural community because of the enlarged range of travel it confers on its owner. The rural population is rapidly making use of the automobile as a means of freight transport, as well as for personal travel and trade. The automobile may be considered at least to double the favored zone around a shipping or market center. Motor trucks have been installed by some of the larger and more progressive farmers, and freight, express and passenger routes have been established between local centers over a circuit of thirty to one hundred miles, at many places in the State. This means of transport affords an efficient and economical method of extending the range of established transportation service to rural communities. Such a line is elastic and need not pass each day over the same route but may cover various routes on different days. Companies are now established to operate these rural automobile lines at different centers as the traffic may warrant. These lines should be encouraged as serving the convenience of rural districts just as the bus and carting lines serve the city dweller and are one of the factors that will aid in making rural life more acceptable.

The telephone has found a large place in rural life and has relieved much of the isolation. In every part of the State, every progressive farmer is supplied with telephone service either by the larger companies or by local coöperative concerns that have a working relation with the large exchanges. In 1914 there were reported to be 2,790,060 miles of telephone wire and 970,449 telephones. This is at the rate of one mile of wire for each three persons and 102 telephones to a thousand of population. The proportion in Pennsylvania was 79, in Illinois 139, and in California 168. Of course, a large proportion of these telephones were in cities but the rural population is fully awake to their advantages and are making an even larger relative use of them than the city resident.

SPECIAL MARKETING AGENCIES

One of the governmental aids to marketing is the requirement that all commission merchants and milk dealers be licensed by the State Department of Farms and Markets in order that their reliability may be assured somewhat. To this provision has been added, in the case of commission dealers in farm products, a fidelity bond of $3000 to guarantee proper remittance for produce sold. There is also provision that the books of a dealer must be kept so as clearly to show the produce received and the disposition made of it, with prices, so that the owner may be assured of honest dealing and the dishonest agent weeded from the great mass of honest ones. There is provision for prompt report of sales on consignment, within forty-eight hours of the consummation of sale. In 1915 there were 252 licensed milk dealers and licensed and bonded commission dealers distributed in the larger cities as follows: New York, Buffalo, Rochester, Syracuse, Albany, Binghamton, Elmira and Olean.

The State has established standards of weight, measure and purity, and in some cases of packages to insure fair dealing and correct understanding between the different agencies that handle farm products. Vinegar must contain 4 per cent acid and 2 per cent of solids. Milk must have 3 per cent fat and $11\frac{1}{2}$ of other solids; butter must not contain more than 10 per cent of water. Standard measures of volume are established and provision is made that products may be bought by weight if so required. Closed packages must be marked correctly with the net weight. Milk bottles are regulated to show the name of the owner and the grade of milk on the cap. Grades are defined for milk and for apples. The standard apple barrel is defined to have a capacity of 100 quarts. The form for labeling many commodities is specified. All these things are in the direction of standardization and honest dealing, and aid in the rapid and direct movement of products at a more stable price.

In 1914 a further step was taken by the State in the establishment of a department or office of foods and markets, later made a division of the Department of Farms and Markets, which is charged with the investigation of market conditions and the dissemination of market information, and is authorized to establish public auction markets in the several centers of distribution and to aid in the organization of farm producers. The idea is that a public auction market under the supervision of the State would be an open court to which producers and buyers may come and at which a fair, open and direct sale price may be established. Not only is this for the mutual benefit of the parties involved but the prices so established when taken with the established grades set a standard by which sales by private agencies may be measured. This auction market under State supervision was first opened in New York City in the fall of 1915. Preliminary to this, the first sales, which dealt with apples, were held in orchards in several parts of the State where the fruit in several adjacent orchards was catalogued in one list for the guidance of buyers. By the extension of these auction markets, by means of a thorough system of bulletins on market conditions, distribution of products and prices from day to day, and with perhaps additional storage facilities to supplement that in private hands, it may be hoped to stabilize prices, make them more satisfactory to producer and consumer and insure a better distribution of the products of the farm. To bring the producer and consumer nearer together, some progress has been made in establishing municipal markets in the larger cities. Extensive investigations have been made of the model municipal market quarters and storage facilities provided by a number of European cities. As a result, there has been strong advocacy of the further extension of this system of municipal aid in the larger cities of the State. The universal interest of the population in the supply, quantity and price of agricultural products is the justification for such public aid and coöperation. In New York City there are public markets where all kinds of produce are sold from the wagon or stall of the producer to the dealer and to the consumer. These markets are poorly equipped and managed and are inadequate to the problem in hand. A municipal committee has been studying the situation and has brought in recommendations looking to the construction of more adequate buildings with storage, at the expense of the city and under its supervision. The cost of such plant would largely be met by charge for selling space in the market.

The city of Rochester has expended $400,000 on an agricultural market building and provides additional space for producer-consumer trade from the wagon. In the building, space is largely occupied by the small dealer or huckster who purchases part or all of his supplies and maintains a regular stand. The market serves as a general exchange for the sale of produce at wholesale, as well as retail.

In Buffalo there are three public retail markets where the producer and huckster have stalls or stands and regularly meet the consumers on the three or more market days of each week. The markets are for both retail and wholesale business.

Some of the smaller cities have established producer-consumer markets. Curb markets are most common, being merely a section of street set aside for this purpose. Others have private stalls. In the latter, both open and closed booths are provided and are rented at a small sum a day or a fixed period depending in part on the size of the stock to be handled. Cities having the former type of market are Olean, Jamestown, Elmira, Auburn, Ithaca and Troy.

These public markets form a link in the system of distribution but can not be relied on to solve marketing problems. They serve as a clearing-house for local produce where the consumers and some buyers may meet the producers. By means of this central meeting place and by the elimination of some delivery charges, prices may be reduced. In practice the operation of these markets is still crude, since the business is a day to day one. There is no adequate information service and no provision for storage except as it may be secured through intermediate dealers. The system of business is rather crude and extremes of supply and demand are not uncommon. The producer's end is a trying one and frequently the volume is so small that the actual cost a unit may be high. The seasonal variation in the products of a single producer, except in vegetables, makes it difficult to maintain continuity of attendance by the farmer which is further increased by the pressure of other lines of work at home at some seasons. The range from which producers are drawn is relatively small and the public market of this sort can not serve more than a small part of the needs of the larger cities. The public market touches only one or two of the elements of a sound marketing system, personal contact between producer and consumer and a sort of local clearing-house arrangement. The nature of the business is such as to be particularly serviceable to the poorer classes who are free to attend and who find grades of produce adapted to their pocket-book that would not be available in such quantity if they had to go through the hands of a series of dealers.

The principle of concentration and volume has not been applied directly to much of the farm produce business by which means the cost of handling is most readily reduced and by which the other elements, such as continuity, advertising and general market knowledge, are best secured. These have not been touched by state or municipal provisions except as some beginnings have been made to encourage coöperation in marketing.

The second means of securing higher marketing efficiency, that of private organization of the producers of a line or group of commodities, may next be considered. These have been slow to develop and have not reached any very large or widespread proportions, but significant beginnings have been made.

In all lines the larger and more aggressive producers have sought to establish direct relations with consumers and efficient distributors in the centers of population. Sometimes, stores have been established, such as milk stores. This has been most practicable for the special or higher grades of product such as certified milk. The best known example in recent years of the efforts of the producers of a single commodity to organize adequately to handle and to secure a fair price for their product is represented by the Dairymen's League. Stimulated by the success of the so-called milk strike for better prices in Chicago in the spring of 1916, and backed by abundant figures collected by a variety of State institutions and investigating committees, the League organization developed to a condition of strength in the fall of 1916, whereby terms of contract could be dictated to the milk dealers of New York City. By subsequent action the principle of collective bargaining between the producer and distributor of milk has been established. The dairymen of many counties were led to withdraw their supply of raw milk and divert it into other channels such as the manufacture of butter and cheese for a sufficient time to show that they meant business. More recently the League has been forced to take steps to care for the surplus of milk at certain seasons and to equalize the price to producers. These two movements, the Chicago and the New York milk strikes, are undoubtedly epoch making in the example they afford of the benefit of united action among farmers when backed by a just cause.

The quantity of milk and other dairy products consumed by New York City is so large that it must be drawn from a region reaching out hundreds of miles into the country. Raw and prepared milk is brought from nearly every part of the State by special milk express trains. They run regularly from the middle of the State and give service as far north as the Canadian line in the St. Lawrence and Champlain valleys. All along the route are milk gathering stations where the product is assembled and prepared for shipment. Some is put in bottles but the greater part goes in forty-quart cans. The manufactured products of milk, butter and cheese, are made up either in privately or coöperatively owned factories. These coöperative organizations have seldom extended beyond a single plant but there are instances in which a series of such plants is operated under a coöperative or corporate arrangement.

A large outlet for milk in many sections of the State where dairying prevails is the milk condensaries and the evaporated milk plants. The former have many large and prosperous plants, but they represent no particular advantage to the individual farmer in determining the price of his product based on cost of production. These manufacturing lines are closely linked with the sale of raw milk for direct consumption. In some of the smaller cities, the retail milk trade is being concentrated and in Ithaca a considerable section of it has been linked with ice and ice-cream manufacture, a modern bottling plant and also a condensing plant. From here the product of a number of dairies is distributed by wagons, each of which is assigned a full, compact and well balanced route. The manufacturing end serves to care for the surplus of raw milk. The business is managed on a corporate basis and is owned and controlled largely by the farmers supplying milk.

Another phase of coöperative marketing is represented by joint sales of pure-bred live-stock. The Liverpool sales near Syracuse have been famous for the amount and high grade of stock handled. By this concentration the owners are able to secure the attendance of a larger number of buyers and to secure more extensive advertising, together with the general advantages of such sales. Other communities hold joint stock sales at a convenient farm with similar advantages. In one instance, Tompkins County, a county breeders' association has been organized that publishes an informational and sales paper to promote the sale of the stock of the breeders in the county and a secretary is paid to pilot buyers to the different farms having stock to sell and to arrange joint sales.

Produce-handling companies comprised of growers have been organized at several places to concentrate, grade and pack their produce, and find a suitable market for the same by means of a paid manager and packing force. The South Shore Growers and Shippers Association, southwest of Buffalo with headquarters at Silver Creek, is an example of this sort of combination of farmers for selling vegetables and small-fruits. Another is the Tompkins County Farmers Company at Ithaca. Still another important concern is the Eastern Fruit and Produce Exchange of Rochester, which was organized primarily to handle fruit in western New York. This corporation, which is owned and officered by producers, is affiliated with a national sales agency with representatives in different cities and doing a business in a large variety of agricultural products. Produce is handled only for affiliated members, which privilege is secured by the ownership of stock. Local exchanges are established in the various centers of production. In an increasing number of places, the local exchanges are establishing central grading and packing-houses to put the produce in a more satisfactory market condition. For example, there were organized in Niagara County in 1918 four central packing-houses for fruit. These selling agencies keep in closer touch with marketing produce conditions than is possible to the average individual; the large volume and variety of produce permit a regular trade to be established and the organization brings the producer and consumer nearer together and on a more equitable economic footing than is possible to individuals working alone. With similar beginnings, the fruit-growers of Nova Scotia and Ontario, Canada, have made large progress. There, the central packing-house has become established and sales agencies have been developed made up of more than forty of these local concerns and handling as much as 460,000 barrels of fruit under a single label.

Still another phase of the distribution is the county farm bureaus. These institutions, financed primarily at public expense and supervised by State agencies, do not and should not take a direct part in the sale of products or the purchase of supplies. But being locally established and in touch with the producers of a county, they serve as the clearing-house for information about farm produce and enable different regions conveniently and directly to get in touch for business purposes. For example, the northern counties of the State produce an excellent quality of potato that is especially suitable for seed for the early crop on Long Island, and by means of the farm bureau officers these two interests have been brought into touch with each other. As it becomes better known, the farm bureau is increasingly being made a means of information touching farm produce and exchange. They also furnish leadership in organization movements as was well illustrated in the New York milk strike.

The other side of the rural trade is represented by the purchase of supplies for the farmer. Of these he is a heavy user and the annual volume of business in these lines is tremendous. Feeds, fertilizer, machinery, twine, household supplies, seeds, lumber, hardware, and a great many other things are used by the farmer. Normally he is a small business man, and, therefore, buys in relatively small lots, often on credit and at a high price. The concentration of his buying, as well as of his selling, brings profitable results. Farmers have been inclined to combine to purchase supplies and there are several buying organizations of farmers where there is one for selling. Perhaps the steps to action appear more simple and the profits more immediate and direct.

The agency that has done the largest business in this line is the Grange. The function of purchasing supplies for its members has from the inception of the organization been a prominent one. This has been effective only in the local centers and was without other help. Sometimes stores are operated and in a large number of the local granges supplies have long been purchased on collective order handled by a member. Frequently this was handled by a member without compensation which entailed an unfair burden, and is conducive of a general looseness in doing business, and may result in its early discontinuance.

Cash trade has been the rule. Many enterprises of this sort have perished from want of efficient management and lack of any overhead support.

As a type of another group of purchasing agencies may be mentioned the Bedford Farmers' Coöperative Association of Mt. Kisco in Westchester County. This is incorporated under the provisions of the general corporation law of the State and has a capital stock of $25,000. It maintains a paid manager and its business is divided into six lines or departments. The development of the several functions of this association very well illustrates the general progress of purchase, marketing and self-help agencies. The expansion of the marketing facilities is one of the last things usually taken up by farmers, yet it is the most vital part of their problem. The purchase of supplies is very much secondary in importance to the effective marketing of products.

Another notable coöperative organization is the Jewish Farmers Association whose activities for its members are not confined to New York State. The latest and most ambitious enterprise for handling farm produce and supplies is an organization known as the Coöperative Grange League Federation Exchange, Incorporated. This is purely a business organization on a state-wide basis. It is an outgrowth of interest in commercial enterprises on behalf of the farmer in three distinct state organizations, namely, the Grange, the Federation of Farm Bureaus, and the Dairymen's League. In addition to an overhead state exchange, it is endeavoring to develop local community organizations with warehousing facilities, in order best to serve the farmers' interests.

A little farther removed from the farm but still directly related to it is the organization of coöperative consumers associations for the purchase of foods and other supplies for the home and household. These organizations form convenient units for dealing directly with farmers' marketing exchanges in the purchase of farm products and are a promising development in the marketing situation. They, with the produce exchanges, form efficient centers on the two sides of the producing and consuming business on the basis of which the intermediate machinery for more efficient interchange may be built.

Two other factors in marketing are storage facilities and credit. In 1911, there is reported to have been condemned in the markets in New York City 73,785 pounds of eggs, 35,755 pounds of fish and 200,000 pounds of poultry. Storage develops at two places primarily, at the centers of production and at the centers of consumption. In New York the various products are cared for by local, usually private, storages, warehouses and elevator cold storage plants at the local shipping stations. In the fruit region in western New York, there are large refrigeration and common storage plants at the main fruit-producing stations, mostly on the New York Central lines. Most of this is said to be owned or controlled by dealers. The storage facilities in the larger cities are gauged to meet the requirements of the minimum amount of produce and the highest practicable price. Adequate storage facilities of a kind adapted to the product are a fundamental essential of efficient marketing in this day of year-long consumption of the staple products of the farm.

The credit factor is closely allied and the storage and marketing conditions reflect the credit status of the average farmer. Having a short margin of cash funds on which to produce his products, they are in most instances marketed as soon as they are matured. The wave of market-movement follows strongly that of production in all the staple products. This means the farmer is selling his crop to cover labor and supply bills that have been contracted, also on a poor credit basis and, therefore, at a high cost. One of the irritating outgrowths of this system is the car shortage in rail transportation that frequently takes an added toll from the farmers' returns and which would be reduced by a longer season for movement.

CHAPTER VIII

RURAL MANUFACTURES OF NEW YORK

RURAL manufacture includes all those manufacturing operations conducted under distinctly farm conditions. In the pioneer days in America, a large part of the manufacturing was thus undertaken on the farm and in the home. Not only simple food products and preserves were made there, but the cloth for the family wardrobe was manufactured from the wool and flax grown on the farm which was carded, spun into yarn and woven, colored and made up in the home. The corn and wheat were ground in the farm mill, which was frequently a custom mill. Lumber was taken from the farm woodlot and sawed in the neighborhood mill for the buildings. The rural family, and especially the rural community, was largely self-sufficient in the manufacture of the things needed by its members.

With the progress of society, including the development of machines and power and the possibility of the concentration of labor in large establishments, there has come about the vast changes in manufacturing processes as they touch the farm and the farm home. A large part of these operations have been transferred from the farm to the city and drawn together in larger and still larger units. This change is strikingly illustrated in the slaughtering and meat-packing industry. In early times, meat was almost exclusively prepared and preserved on the farm, and the social "bee" was a common adjunct. Now the business is mainly concentrated in a few large packing centers, such as Chicago, Kansas City and Buffalo. Butter, instead of being made on the farm, is now largely manufactured in creameries or butter factories that handle milk by the thousands instead of by the tens of pounds.

As a result of these changes, it is not possible to draw a sharp line of distinction between urban and rural manufactures. The distinction is one of size rather than character. Some industries are still in the transitional stage and have both a distinctly rural and an urban aspect. Vinegar-making is one of these. It is still made on the farm from cider extracted at the local custom mill, but it is also manufactured in city factories far from the rounds of the farm. Even in New York City vinegar is made, whether from apples does not appear. Indeed most forms of manufacture began on the farm and certainly in close contact with the home, and grew beyond its limits with the changed facilities of new times.

The history of these manufactures is a fascinating chapter in human development and there is strong temptation to trace some of the features of it. The history is not peculiar to New York, however, and space will not allow the digression. The reader will know in a general way what the changes are like.

The "drying" of fruit is an illustration. In the early days the fruits were dried about the kitchen stove on sieves and wooden lattice-shelves. The trays were set out-of-doors in sunny weather. Often the halves and quarters of apples were run on strings by means of a darning-needle, and the strings were hung about the stove and ceiling; the sanitary results are left to the imagination of the reader. In regions of long absence of rain, as in California, the sun-drying of fruit soon came to be a common and good commercial practice. Later came the "evaporated" fruit, when the prepared product was placed in a tower or specially constructed building (an "evaporator") in which the moisture was driven off by furnace-heat or steam-heat. Vegetables and many other products are now preserved by "dessicating" and "dehydrating," in which the mechanical processes are still further perfected. All these classes of industries have developed strongly in New York.

The days of "home-spun" are past, and the farmer now buys most of his supplies where other persons buy them. The traveling craftsman has gone. The farmer raises his products for sale rather than merely to supply his own needs, and he is a heavy buyer as well as a producer. Something of the picturesqueness of country life has left it with the passing of the local tannery, grist-mill, wagon-shop, broom-shop, barrel-factory, hand-loom, shoemaker's-shop and cabinet-shop, but the rural people may the better concentrate themselves on production.

Attention in this chapter will not be confined to

purely rural manufactures but will survey briefly those forms of manufacture, whether rural or urban, that directly use the products of the farm and put it into some new or more convenient form to use. It will be noted so far as possible the extent to which these industries have been transferred from the rural district to the larger centers.

It should be noted that New York is the leading manufacturing state and is situated near the center of the main industrial region of the country. This follows naturally from the concentration of population. In 1909 about one-sixth of the total value of manufactured products in the United States was supplied by New York with a total value of $3,300,000,000. The figures that follow are generally for the last census year of 1909. The value added to the product by manufacturing operations in New York was about $1,500,000,000 or $1,250 for each worker. Of this amount, the workers receive $625; $125 may be assigned for interest, leaving $500 to a worker, distributed to other interests. Herein lies one of the problems of distribution which will receive increasing attention in the ensuing years. There should, of course, be a legitimate margin for the agencies of manufacture but it is questionable whether an adequate basis for gauging that margin has been applied. Clothing, food-stuffs and timber products are the largest items in manufacturing enterprises. This is of significance in its relation to agriculture since it constitutes an expansion of the local market as represented by population and local and seasonal consumption. With the exception of Pennsylvania, New York State has the largest diversity of manufacturers. Out of 264 classifications used in the thirteenth census, 243 were represented in New York.

The materials of manufacture are the products of the farm, the forest, the stream, the mine and the quarry and somewhat of the sea. The development of manufacture in the region represents largely the persistence of early established industries, together with the transfer of others as a result of changed natural advantages. Tobacco manufacture exemplifies the former. Now widely disseminated, it shows relation to the former large production of tobacco in certain parts of the State. In the case of meats and the meat-packing industry, there is recognizable the tendency to transfer towards the centers of population from the districts of production.

Dairy manufactures were once a conspicuous feature of the farm operations. Butter and cheese were made on the farm and certain regions became widely known for their product, especially western New York and the district around the eastern end of Lake Ontario. The total production of milk in the State in 1909 was 2,400,000,000 quarts or 60,000,000 forty-quart cans. This number of cans would fill 240,000 forty-foot cars, enough to reach from New York to Chicago in a solid line. They would make 12,000 trains of twenty cars each or an average of thirty-three a day to move the milk daily produced in the State. Seven-eighths of the milk was sold as such. However, a considerable part of the total milk goes into manufactured products. Dairy manufacture is still distinctly rural but it is undertaken in central factories rather than on the farm. A little more than one-third of this amount is manufactured in butter, cheese and condensed milk in factories off the farm. There is relatively little butter made on the farm. The manufacture of cheese especially has been transferred from the farm to the factory. In addition, other forms of dairy manufacture have been introduced, such as condensed and evaporated milk and casein. In 1909 there were 1,552 factories handling milk. Of these, 426 were engaged primarily in the manufacture of butter, 1,090 in cheese and 36 in the manufacture of condensed and evaporated milk. Their products represented the following proportions of the total dairy manufacturing industry: butter 41, cheese 36.6, and condensed milk 23.4 per cent.

The distribution of the factories making these products follows closely that of dairy cows, but in the main it is pushed off to the more remote districts where market milk is not sold. The sale of raw milk for direct consumption has been rapidly increasing for twenty-five years. New York City now consumes about 22,000,000 cans (forty quarts) of milk annually, equivalent to about 180 to an inhabitant. To meet this tremendous demand, the milk-handling facilities have reached out in all directions hundreds of miles along the railroads until milk is now shipped to New York City from the remotest corners of the State, from far into western New York, from the St. Lawrence Valley, from the upper Champlain and even from over the Canadian boundary. Milk express trains run into the city daily on all the important railroads. Naturally this drift to New York City is deflected around the smaller towns and cities which in their turn are reaching further and further for their supply of raw milk and contribute to the general trend away from the manufacture of milk products.

Coincident with this large development of the business in raw milk and cream, there has been a tightening of the sanitary standards under which milk may be made and handled. Such regulations now cover inspection of the barns and cattle, tests of the cattle for tuberculosis, grading of milk, provisions for cooling and pasteurization, for types of containers and regulations regarding cleanliness. This is reflected in the market grade of milk.

A further development in the dairy business is the production of certified milk, particularly for infants and invalids. The regulations are particularly strict and the price must be correspondingly higher than for ordinary milk. The beginning of the certified milk business was about 1900. In 1910, twenty-nine farms produced certified milk to the amount of 16,500 quarts a day.

The total production of milk has decreased about 23 per cent in the ten years from 1899. Since 1910 it has increased slightly. While the sale of market milk has increased 18 per cent, the production of butter has decreased nearly 40 per cent which is part of a continuous decrease for thirty-five years from 1880 when the maximum production occurred. The manufacture of cheese has decreased 60 per cent in ten years which is part of a continuous decrease for sixty-five years from 1850. This antedated the development of machines and equipment for the manufacture of butter in factories, chief of which was the milk separator that was invented in 1879 and came into general use in the eighties. Butter and cheese are less perishable than raw milk and cream and their production is relatively more economical in regions remote from market, where stock feed is cheapest. Market milk must be produced as close as possible to the center of consumption.

The total production of butter in 1909 was 69,358,918 pounds, valued at $13,500,000. The rank of New York was eighth, Wisconsin leading as also in the production of cheese. Of this amount, 66 per cent was made in factories and 34 per cent in farm dairies.

The first creamery or butter factory in America was operated at Campbell Hall, Orange County, in 1856. That county was the first important dairy center and "Goshen butter" was a standard market grade in the middle of the last century. In 1914 there were 576 creameries in the State, Minnesota leading in the United States with 848 factories and Wisconsin second with 812.

Among the very earliest American factories of milk separators and other dairy supplies was one at

Little Falls. The Babcock milk test, named after its inventor, had an early connection with New York through the fact that Dr. Babcock, its inventor, was born and trained in the State and later worked at the Agricultural Experiment Station at Geneva.

The production of cheese has gone much further than that of butter in the transfer from the farm to the factory. Over 99 per cent of the cheese in 1910 was made in factories. The total production was 105,584,947 pounds, valued at $14,250,000 in 1909. The making of cheese is likely to continue to decline. The distribution of cheese factories is much more bunched than is the creameries. In 1914 there were 995 factories, placing the State second in rank in number. New York is also second in total production of cheese. Wisconsin is first with 1720 factories.

The two centers of cheese production, of nearly equal size, are the five southwest counties and a broad belt reaching southward from the Canadian line in the St. Lawrence Valley to near the southern boundary of the State by way of the Chenango and the Unadilla valleys. There are very few cheese factories outside of those regions. The two largest market centers for cheese are Watertown, in Jefferson County, and Cuba in Allegany County.

To the uninformed, the word cheese usually means the common American Cheddar or hard cheese. The history of the development of this product is very thoroughly wrapped up with the history of dairying in New York. The rise of the industry was based on the difficulty of making butter in the hot weather of summer and is coupled with the practice of summer dairying. The development of the industry reaches back to 1804 when Colonel Jared Thayer, of the town of Norway, Herkimer County, made cheese from the milk of a herd of twenty cows. Many details of manufacture and curing had to be overcome, as suitable rennet, proper heating, cutting and grinding the curd, size of cheese, the relation of butter-fat to yield, and finally of factory as compared with farm manufacture.

The need for a larger and more uniform cheese made it necessary to handle a larger quantity of milk in one lot than the average herd provides. This gave rise to coöperation among farmers to increase the bulk of milk. About 1851, Jesse Williams, of Oneida County, combined the milk from his sons' farms with his own. This was the germ of the coöperative factory system destined to have so large a rise. In 1879, New York produced 56.5 per cent of all the cheese in the United States and 33 per cent in 1910.

While the bulk of the cheese is of the American Cheddar type, it is surprising how many kinds of cheese are made. With many of these forms the ordinary person has no acquaintance since they are not staple market products. The last available summary of these was in 1908 when nineteen different kinds are reported and the production of several others is indicated. In Table XIII is given a list of these and the amount of each kind produced:

TABLE XIII.— KINDS AND AMOUNT OF CHEESE PRODUCED IN NEW YORK IN 1908

1.	American Cheddar	77,821,109 pounds
2.	Skim	10,725,699 pounds
3.	Pineapple	255,778 pounds
4.	Limburger	6,187,801 pounds
5.	Domestic Swiss	968,952 pounds
6.	Kosher	213,959 pounds
7.	D'Isigny	978,454 pounds
8.	Neufchatel	1,905,263 pounds
9.	Framage de Brie	172,660 pounds
10.	Squore Cream	1,147,442 pounds
11.	Imitation English Dairy	337,310 pounds
12.	Weiner	4,800 pounds
13.	Sage	180,589 pounds
14.	Munster	671,881 pounds
15.	Pot	7,639,364 pounds
16.	Italian	1,459,144 pounds
17.	Pressed	439,092 pounds
18.	Casciocavalio	354,421 pounds
19.	Various other kinds	782,686 pounds

In the manufacture of cheese of the special or fancy grades in particular, there seems to be an opportunity to supply a most desirable food product and to expand the demand and correspondingly improve the price of all milk. A taste for these special brands of cheese can be very readily cultivated.

Of condensed milk New York produced 120,500,000 pounds worth $9,500,000, and ranked first. Of casein the State manufactured over 6,500,000 pounds, worth nearly $500,000 and again had first rank.

The production of condensed milk was equivalent to about 12 per cent of the total milk in the State, and nearly 30 per cent of the total amount in the country. The factories are widely distributed in the dairy regions, particularly in those parts remote from the larger cities, and usually in connection with plants that make a business of shipping raw milk. Nearly three-fourths was sweetened and one-fourth unsweetened. The plants require rather expensive equipments and are generally established in fairly large units. For the production of milk powder there were in 1918 eight plants in the State which increased in 1919 to twenty. In 1918 they produced 2,398,849 pounds of whole milk, 524,873 pounds cream and 11,531,487 pounds of skimmed-milk.

Farmers still butcher some animals and sell the carcasses. In 1909 this constituted only 8 per cent of the total value of animals slaughtered in the State. There is also a little butchering of animals purchased by retail dealers in their region, but the aggregate of this business is relatively small.

The preparation of cured meat has largely moved from the farm to the city factory. The value of animals slaughtered on farms amounted to $9,927,000 in 1909, which was $50 a farm or approximately the value of two fat pigs, or a small beef. It is evident that this would be consumed largely on the farm, leaving very little for sale in the cities. The total value of meat slaughtered and packed in the State was $127,130,000, not including that killed on farms or by retail butchers. This figure places New York third in the list of states, with Illinois and Kansas, first and second. The former handles considerably more than the sum of the other two states. These figures illustrate very well the concentration of finished animals in the larger centers of population for slaughter. The value of animal products exclusive of the dairy is out of all proportion to the number of meat animals in the State. Undoubtedly, the figures represent the marketing of farm products in this convenient form because of the advantages of shipment. In New York the bulk of the slaughtering and meat-packing industry is in Buffalo and New York City which had in 1909 an aggregate value of $121,290,000, which is 95 per cent of the total. Over 80 per cent of the persons engaged in the industry were in those two cities. Curiously enough nearly four-fifths of this industry in the two cities is reported from New York City. In both cities this production represents a material increase in the last ten-year period in spite of the general decrease in meat cattle in the State. The three minor cities of Rochester, Utica and Albany together had an aggregate value of only $591,000.

The distribution of the values of the main products of the slaughtering industry is as follows:

TABLE XIV.— VALUES OF MEATS

Fresh beef	$41,428,480
Fresh veal	4,316,483
Fresh mutton	9,539,554
Fresh pork	12,665,175
Salted pork	2,686,225
Hams	7,053,027
Shoulders and sides	8,064,812
Lard	5,814,101
Hides	7,105,591
Sheep pelts	1,389,471
	$100,062,919

This represents the production of 238 plants that employed 6,110 persons and added about $16,962,-000 to the total. The economy in large centralized plants is especially well exemplified in the meat-packing industry. The possibility for the utilization of by-products has been realized. A large branch of the fertilizer industry has grown out of it. The manufacture of soap, glue and oils, the utilization of hides, bones, horns, hair and every conceivable part of the animal is made possible in these large establishments. Formerly they were wasted or poorly utilized in the small packing-house.

The local tannery used to be a feature of every considerable community, but now it is often difficult to find a local buyer of hides. New York stands fifty in the leather and tanning industry with a value of $27,642,000. This is an increase of 16 per cent for the preceding ten-year period in comparison with an increase of 56 per cent in the slaughtering and meat-packing industry. The value of the untreated hides produced in the State was $8,500,-000. The production of hides doubled in the ten-year period preceding 1909, probably representing considerable importation of foreign hides.

New York produces nearly 23 per cent of the soap of the country, nearly all in New York City and Buffalo in alliance with the meat-packing industry.

Sixty per cent of the value of gloves in the United States is produced in New York, largely localized in a small region in the Mohawk Valley where the town of Gloversville marks its center. This industry

Plate VII. Typical view in the hill country of Southern New York. Dairying is the leading industry.

has a distinctly rural aspect since a considerable part is piece work done at home in farming districts.

Of boots and shoes, the State produces about 10 per cent of the country's product with the region of Binghamton and the city of Rochester preëminent in the business.

There are fifteen fertilizer plants in the State, most of which are based on the utilization of animal wastes and they are closely allied with the meat-packing and rendering establishments. To these waste products are added the phosphate and potash materials to make a complete fertilizer. The larger plants are equipped to treat the raw phosphate rock with acid. These fifteen plants with a capital of a little over $3,500,000 produced goods to the value of $4,250,000 in 1909. The concentration in this business, as in others, is illustrated by the fact that in 1899 there were thirty-two plants with a capital of over $4,500,000, but with a product worth only $3,100,100 without any decided change in the value of materials. The bulk of the plants is in Buffalo and New York but small ones are at Newburg, Albany, Fulton, Utica, Syracuse and Rochester, with minor plants at other points concerned with special materials, such as wood-ashes and wool waste.

The prominence of New York in the lumber industry has already been noted in the discussion of forest resources. For several decades the State has been decreasing in the production of lumber. This decrease amounted to 72.4 per cent in the last ten-year census period. At the same time, the leading varieties of lumber cut have changed and the poorer qualities have attained relatively greater importance. Hemlock leads, with spruce and white pine next in order. In spite of the large decrease in timber products, the planing mills contribute 65.9 per cent of the product in the country from 674 plants. New York furnished 20.7 per cent of the product of saw-mill and logging plants with 1,389 establishments and 200 packing-box factories contributed 13.4 per cent of the business of that industry. The customs and portable saw-mill adapted to cut-up farm tracts of timber is still a common sight, there being 211 establishments. They employ only 452 persons and have a capital of less than $500,000, their product being valued at less than $250,000.

The cooperage business still has a distinctly rural aspect, there being many small plants in the apple-producing region. Many of them operate only in the fruit season. Here, also, the tendency is to transfer the industry to the larger concerns and machine-made barrels are taking the place of the hand-made ones. The bulk of the cooperage product is made in the larger cities, New York with thirty-five plants contributing 45 per cent, Rochester 9 and Buffalo 8 per cent. The total value of the product, including some secondary forms of wood work, was $7,500,000. Doubtless a large part of these packages was required for the products of manufacture such as sugar in New York and flour in Rochester. There were 364 cooperage and other wood-working plants in the State employing an average of nine persons each.

Elm and ash were the local woods used for fruit and flour barrels, while the gums for the heads must be imported from the South. Oak for casks may still be secured locally.

New York produced only 15 per cent of the wood pulp made in the United States. This came almost entirely from spruce. The twenty-one plants engaged exclusively in this industry and the fifty-one factories that made paper in addition, produced 314,000 tons, valued at a little less than $10,000,000. The State uses great quantities of paper of all grades, about one-fifth of which is made up locally but largely from imported materials. The mills are mostly around the base of the Adirondack Mountains where water power is available for reducing the wood to pulp. Every important city in that territory contributes to this industry and plants are located on the most important waterways leading from the mountains.

In baskets, rattan and willow ware, New York, with 30 per cent of the workers and 37 per cent of the plants, produces 60 per cent of the value of the products in the country. There were 169 establishments for this purpose. The statistics do not permit an estimate of the use of local materials, particularly hickory, ash and willow. The preëminence of the State in growing willows in the Wayne County district undoubtedly contributes something to this line of manufacturing and gives it a local rural aspect.

One of the most distinctly rural manufactured products is maple sirup and sugar. The sugar-house, with its ventilating cupola, is usually a small frame structure where the sap is collected and evaporated to sirup or sugar. It is interesting to note that there was a substantial increase in the production of maple sirup and sugar in the United States in 1909 over that in 1899. The same statement applies to New York, which also produced more maple sugar and sirup in 1918 than in 1910. In 1918, the production was 3,732 pounds of sugar and 1,755,000 gallons of sirup. New York leads in the production with Vermont a close second, and the two states produce 75 per cent of that made in the United States. Sirup and sugar-making is the first activity on the farm in the spring, beginning when clear warm days and sharp frosty nights set the sap running and the buds begin to spring.

Cider and vinegar are largely rural manufactures of which the State produced $2,250,000 worth, or over 60 per cent of the total of the country. Vinegar is becoming more and more a centralized instead of a secondary farm product, as was once the case. Cider remains essentially local, made in custom mills scattered wherever apples are grown. As cider for direct consumption, it is not suited to distant transportation. Nearly 5,250,000 gallons of cider are reported from farms in 1909, an increase of nearly 1,000,000 over the preceding decade. Twenty-five per cent of all farms reported cider as a product. Much of the custom-made cider returns to the farm and passes on into vinegar under the simplest possible manipulation. While some of this is sold, it is available in such small quantities and of such variable quality that dealers and grocers prefer the factory-made, standardized product, the grade of which comes under the inspection of the State. To extend the use of apples the United States Department of Agriculture has devised a product known as apple juice suitable as an unfermented beverage, which is rapidly attaining a place of prominence. The inception of prohibition has given a strong impetus to the manufacture of pasteurized cider. In 1909, 277 plants were engaged in the manufacture of vinegar and cider. Of these four were in New York City but their product represented only 6 per cent of the total value.

The fermented product of the grape, wine and its derivations, has been decreasing in production for a period of years but the unfermented product, grape-juice, has offset this decrease. The Westfield factory, which was the pioneer in this business, has been followed by many others, including factories in the grape-producing section of the Hudson Valley opposite Poughkeepsie. The development of the unfermented juice industry began in Chautauqua County about fifteen years ago and over 3,000,000 gallons are now produced annually in that district and represent a substantial outlet for fresh grapes. Sparkling wine, commonly known as champagne, has continued until very recently to be produced in all the important grape-growing centers. Seventy-five per cent of the champagne in the country has been produced in New York and with grape-juice has been estimated to consume a quarter of the gross tonnage of fruit.

Dried and canned fruits are also an important extension of the fruit industry. Including vegetables, New York is the ranking state in these products. Seven hundred and ninety plants produced $19,000,000 worth or 23 per cent of the total. The materials used for this purpose and valued at $11,500,000 represented nearly one-third of the value of such fruits in the State. California leads in the production of dried fruits. Of preserves and pickles the State yields $7,000,000 in value, equivalent to 15½ per cent of the total for the country.

Vegetables are canned in plants of medium to large size distributed for the most part in rural districts in the vegetable-producing region of central New York, particularly the territory from south of Utica to near Batavia. Peas, corn and beans in this order of value make up the bulk of the product. In the same region, where cabbage is most largely produced, the manufacture of krout is also centered, particularly from near Syracuse to Batavia.

An even wider range of fruits is canned, embracing in the order of value, apples, berries, cherries, pears, peaches and plums. Apples make up more than a third of the total. These plants are generally in close proximity to the producing centers for those fruits.

In dried fruits, apples constitute 90 per cent of the value and represent the chief means outside of cider and vinegar of utilizing the poorer grades of fruit. The plants are closely identified with the apple orchards and are generally located in the smaller towns. The remaining percentage is made up of small-fruits, chiefly raspberries, which centers in Yates County on the west side of Seneca Lake.

The preserving industry, on the other hand, is developed in larger cities, particularly in New York City.

New York State ranks first in the quantity of corn, buckwheat and oats ground, third in that of wheat and rye, and sixth in barley. There were 983 mills in 1909 of which 270 were engaged in milling wheat, and of these 105 produced less than 1000 barrels during the year; 86 produced from 1000 to 5000 barrels; 49 from 5000 to 20,000 barrels; 19 from 20,000 to 100,000 barrels; and 11 plants produced over 100,000 barrels. These figures give some idea of the concentration of milling in extensive plants in large cities, particularly Buffalo and Rochester.

Milling illustrates the relation of manufacturing to production and special industries. New York leads all others in buckwheat and this grain is ground to flour in a larger number of small mills than is possessed by any other state. These often use small water power, particularly in the southern tier counties. Corn and oats from the Mississippi Valley region are purchased to serve as feed for the large dairy and live-stock industry of New York for which purpose they are extensively ground together in these smaller mills. New York was once the leading wheat-producing state and Rochester the chief wheat market and milling center. With the change in the producing center, Buffalo has gained the position as a shipping terminal.

With the saw-mill, the grist-mill was among the earliest rural factories, the abundant water power of the State being favorable for both. Both the first grist-mill and the first saw-mill operated in America were established in the new settlement of New Amsterdam on Manhattan Island soon after it was founded by the Dutch in 1623.

The smaller mills have more and more felt the shift in conditions and the competition of the larger mills, and many of them have been abandoned.

Any summary of the possibilities of agriculture in New York, assuming the basis of soils and climate, must take account of the development of manufactures based on the products of the farm. While the tendency will always be to prefer sale for immediate consumption, the manufacturing processes continually expand the range and period of consumption and, together with improved storage facilities, tend more and more to expand the season so that the products of the farm are conveniently available the year round and at a more uniform and stable price both to producer and consumer than would otherwise be possible.

CHAPTER IX

THE ADMINISTRATIVE AND REGULATORY ORGANIZATIONS OF NEW YORK

GOVERNMENT increasingly touches the individual as the development of the State progresses. While in some respects it places restrictions on his actions and movements, at the same time it gives him a larger measure of freedom and a higher degree of efficiency. The true function of the State, using this term in a broad way to represent all governmental organizations, is sometimes lost sight of, and it is conceived to be a sort of impersonal overlord to be exploited by some and evaded by others, according to their individual situation. The one breeds the mere political heeler and the privileged interests. The other develops the poacher, the short weighers, the profiteers, and the intriguers of all sorts.

Government is a combination of institutions to do for individuals singly and collectively what they alone can not accomplish so well. This is the real governmental function to be kept steadily in mind as the ideal and is the principle that should guide in testing each new idea in legislation.

Strictly speaking, there is no sharp distinction of interests between city and country. They are only a little different in quality, and the best interest of one is in the end for the welfare of the other. Every example of coöperation and of friendly understanding between city and country is, therefore, welcomed. There is no reason why the farmer, as well as the merchant and the manufacturer, should not be a member of chambers of commerce and other commercial organizations centering in the city and be chosen to public positions of trust and honor if he has the ability to hold them.

The principle of responsible representative government should give to State or other large units the administration of the affairs that concern all the individuals alike, and should reserve to the subdivisions those that involve only smaller groups of individuals. Trouble always arises when either attempts to override the other.

POLITICAL DIVISIONS AND SYSTEM OF TAXATION

New York State is subdivided into sixty-one counties, nine hundred thirty-two townships and something over twelve thousand school districts (12,135). These are the main legislative and administrative units of the State and are combined in various ways for special purposes. For example, the counties are grouped into forty-three Congressional districts, and into fifty-one State Senatorial districts. In addition there are village and city units that frequently are set off from the minor divisions or overspread and supplant them.

Government touches the farmer in establishing

standards for the measurement of his products, in regulating the conditions of their production and sale, in the administration of protective measures against the enemies of plant and animal life, in the provision of roads and other transportable facilities, in the regulation and conservation of the natural resources of the country and in the dissemination of standard and current information about all these questions. Thus, a great body of governmental organizations and institutions dealing with agricultural matters has grown up. All must have funds and must be supported out of the means of the people of the State. Taxes are the support thus required and the methods of raising revenues is of prime concern to all.

The total assessed value of the real property in New York heads the list of states and amounted in 1912 to about $12,000,000,000. Approximately $2,000,000,000 is assigned to rural property and includes with the open country all cities and villages having a population of less than 2500. In 1912, the total wealth of the State in all forms of property was estimated at about $25,000,000,000, which was a little more than one-eighth of the total value of all property in the United States. The revenues of the State are derived less from a general property tax than from other and less direct sources of income. In fact, in 1912 only about one-sixth of the total revenues were derived from a direct tax. The largest single item in the State's revenues is the inheritance tax. Other important sources of revenue have been liquor licenses, the series of general and special corporation taxes and a number of miscellaneous items, largest among which are the stock transfer tax, mortgage tax and motor vehicle tax. There is no personal property tax.

There is but one assessment roll for property. For revenue purposes the assessment is supposed to be at the full value, by which is meant the amount of money the property would sell for at a fair, free and well advertised sale. As a matter of fact, this is not the case, and in 1912 the assessed value of all real property was estimated to be approximately two-thirds of the true value. With respect to farm lands, the rule generally holds that the better the land the lower is the assessment relative to the actual selling value of the land, and the poorer the land the higher is the assessed value. Probably, this condition is due to the general lag of one value in following the other. Farmers in the more remote and poorer regions where sales are slow often report that land is assessed for more than it would bring at a public sale.

The average value of farm land including buildings as determined by the census of 1909 ranged from less than $10 in thirty-four counties, embracing all the higher and poorer parts of the State, to more than $125 near New York City, and $60 to $100 for the land adjacent to the larger cities and embracing, in the main, the lower lying valley and plains lands, the average being made by counties. From a study of the assessed value of townships that are known to be strictly rural, it has been determined that the poorer farm lands that are on the border of profitable utility are assessed at a value of $6 to $15 an acre on the average. The bulk of good farm land reasonably well situated is assessed at a value of $30 to $80, while land assessed at a higher value is more favorably located or devoted to special purposes. The total tax rate ranges from $1 to $8 a thousand. The great bulk of the land is assessed at a rate ranging from $1.50 to $2.25 a thousand.

RURAL CREDIT FACILITIES

A state law enacted in 1914, that may have considerable value to the farmer, is the arrangement by which the united credit of a group of individuals may be pooled for the benefit of persons in the group that may need financial aid. It has sometimes appeared that the farmer was at a disadvantage in securing funds and credit to carry on his affairs, and that these facilities were not available to him with the same freedom that they may be had by the business man in the city. Unquestionably this condition has prevailed to a degree, not because there was any desire to discriminate against the farmer but because his business has often not been conducted in such a was as to insure stability in the values back of his credit. As a matter of fact, the reason for much of this uncertainty and even the loss that has resulted has been in the general downward shift in rural values that has prevailed in eastern states during the last forty years. When to this is added the complexity of the factors that make up rural values and the lack of standardized and often of an organized system of business management, the city bankers, unfamiliar with such conditions, were often warranted in their antipathy toward rural loans on anything except the very best farm real estate. The farmers on the poorer lands have always experienced these financial difficulties most acutely. It is a phenomenon associated with depressing rather than with expanding values.

Credit unions may now be formed by a group of individuals under the supervision of the State Superintendent of Banks with shares of stock of a par value not to exceed $25, to liquify and make available the rural credit facilities. Loans to exceed $50 must be secured in a manner to meet the approval of the loan committee. Such a group is in a better position to deal with the larger banking institutions than is the average individual. The farmer with his income often available only at long intervals, and with the necessity of making long-time investments in equipment and improvement, is especially in need of the best sort of credit facilities.

The Federal Farm Loan Bank is now more utilized for the provision of coöperative rural credit than is the Credit Union system. This law was passed in 1916, and New York State is administered from the regional bank at Springfield, Massachusetts. Under this plan, a group of farmers may form a federal farm loan association and become stockholders in the stock of the Federal Farm Loan Bank. The minimum aggregate amount of loans with which such a local association may start is $20,000 and this sum must be divided among at least ten applicants. Loans may not be placed in an amount exceeding 50 per cent of the appraised value of the property by which these loans are secured. A loan once placed is regularly amortized on a basis of 1 per cent a year out of the gross 6.0 or 6½ per cent a year which the borrower now pays on his loan. Up to March 1, 1920, the total amount of loans in this form in New York State was $5,048,440.

THE COUNCIL OF FARMS AND MARKETS

(See Fig. 34)

The Council of Farms and Markets is administered broadly by a council of eleven men, of whom ten are appointed by the Legislature, one to represent each of the nine judicial districts and one by the Governor at large. The Commissioner of Markets of the City of New York is ex-officio a member of the Council.

This Council came into existence in 1917 as the successor to the two independent Departments of Agriculture and Foods and Markets which now constitute the two divisions under the Council. In addition, there is a Bureau of Accounts and a Legal Bureau, each of which functions for both divisions of the Council in matters touching the interpretation of agricultural laws and regulations, and in the prosecution of violations of these requirements. It works in close coöperation with the office of the Attorney-General. Each division is administered by a commissioner chosen by the Council who acts within its general policies and directions.

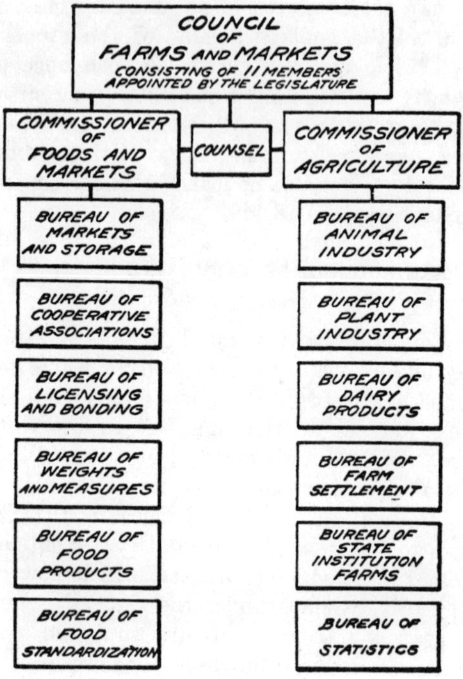

Fig. 34. Diagram showing the organization of the Council of Farms and Markets into divisions and bureaus.

The Division of Agriculture has the responsibility to promote agriculture, especially in a regulatory and statistical way, and to enforce the agricultural law. The function of the Division of Foods and Markets is to facilitate marketing and distribution and to enforce laws establishing standard weights and measures and regulations relating to pure and wholesome foods.

For purposes of administration, the Division of Agriculture is divided into six bureaus outside of the central administrative office. These are Animal Industry, Plant Industry, Dairy, Farm Settlement, Statistics and State Institution Farms.

The Bureau of Animal Industry handles all matters touching animal production that come within the province of the Council. It has to do with the eradication of infectious diseases such as tuberculosis. Regulations for any type of infectious disease may be made and enforced by this office. On request of owners, animals are tested for tuberculosis. Milk from tubercular cows cannot be used except under certain prescribed conditions. Owners of herds are not required by State Law to have tuberculin test of milch cows, but cities may make such requirements as the basis of the privilege of selling milk in a city.

Animals afflicted with infectious diseases that are killed under the direction of the State are paid for according to a scale of prices based on the appraised value of the animal. The limit of value for unregistered animals is $75, and for registered ones $125. Of these amounts, 90 per cent will be paid for animals having the disease generalized and 25 per cent for those having it localized. This applies particularly to tuberculosis. Among the outbreaks of acute infectious diseases that have been repressed in recent years are the foot-and-mouth disease of cattle, glanders in horses and cattle, rabies in dogs and cholera in hogs. This Bureau encourages the upbuilding of the animal population of the State, enforces the stallion enrollment law, and the law providing for the fostering of sheep through the licensing of dogs and the remuneration of owners of sheep for damage done by dogs.

The Bureau of Dairy Industry has charge of the enforcement of the agricultural law relating to dairy products, including butter, cheese, condensed milk and oleomargarine, the use of cheese brands, protection, care and handling of milk, care and feed of cows from which milk is produced and sold in the markets of the State, and the use of suitable receptacles to the end that all milk may be clean and healthful. Artificial butter such as oleomargarine and imitation butter may be sold only when it is not colored in imitation and is clearly labeled as to its character. A certain amount of educational work is undertaken among the operators of butter and cheese factories and in the examination of applicants desiring license to test milk and cream by the Babcock method.

This Bureau also administers the fertilizer and agricultural lime inspection laws. A staff of inspectors collect samples of material wherever it may be found in the State. These are taken in triplicate and one set is forwarded to the State Agricultural Experiment Station at Geneva where the analytical work is done.

The Bureau of Plant Industry bears the same relation to crop production and protection that the Animal Industry and Dairy Bureaus bear to animal production and products. It is concerned with nursery and nursery stock and orchard inspection for the repression of destructive plant diseases and insect enemies. No person is permitted to keep any plant affected with certain infectious diseases such as yellows and little-peach of the peach and black-knot of the plum. Nurseries are inspected and given certificates of health. More than five hundred such certificates are issued annually. Nursery stock coming into the State must be inspected at its destination and carriers are required to give the department notice of shipment. Foreign consignments are inspected at the port of entry and the customs house brokers handling such stock are licensed.

Among other provisions of law is one prohibiting spraying fruit-trees when in full bloom in order to protect honey bees. The inspectors of this office look after the trouble known as foul-brood in bees which is very destructive of colonies. The office also enforces all regulations concerning spray materials for the control of disease and insect enemies, and issues certificates for each brand. Spray materials are sampled and analyzed in the same way as feeds and fertilizers. The cost of a large part of this inspection is borne by those benefited, by means of a license or tag tax and by other assessments. The San José scale, European corn root-borer, gipsy and browntail moth and the blister-rust of the white pine have been particular objects of attention in recent years.

This Bureau also has the administration of the apple grading and packing law. Apples are divided into the following grades: fancy, A grade, B grade and ungraded, and all closed packages must be labeled in accordance with the definitions of these grades. This includes standards of size. The size of standard packages is defined.

The Bureau of Farm Settlement aids in establishing settlers on farms to increase food production and for this purpose has published annually a large bulletin giving a list of farms for sale with a description of each. It also assists in finding places for laborers desiring farm work, especially those from foreign countries. Its chief activities have centered around New York City where many persons have been placed.

The Bureau of Statistics collects all statistics concerning agriculture and agricultural production in all its branches. This is embodied in State census reports from time to time. It aims to keep posted on the condition of farm lands as to value and production.

The Bureau of State Institution Farms has the supervision of the management of the farms connected with forty-two different State institutions, the area of some of which aggregates several hundred acres. Some of these are large business enterprises and in total their product contributes very considerably to the support of the inmates, many of whom are employed in the work of such farms.

The Division of Foods and Markets bears a more

direct relation to the consumers of farm products than to farmers. Outside of the central administrative office, it is divided into six bureaus: Markets and Storage; Coöperative Associations; Weights and Measures; Licenses; and Food Standardization and Food Products.

The Bureau of Markets and Storage has for its special province the giving of assistance to farmers of the State in marketing their crops and the making of arrangements for bringing food from producers to consumers in the most efficient and economical way possible. In each instance, close coöperation between producers, shippers, manufacturers, wholesalers, retailers, and consumers is being encouraged.

For the farmer, a market news bulletin is maintained, showing daily prices and conditions obtaining on the New York City and Buffalo wholesale markets. This is sent regularly to farmers on request. There has also been established a Grievance and Complaint Service to which farmers and shippers of produce are invited to send in, for investigation and speedy adjustment, any data they may have showing a lack of honorable and efficient service on the part of distributers.

In the interest of equalizing distribution and controlling of market fluctuations, the Bureau is taking steps to establish standard grades for various kinds of produce and to bring about the licensing of all distributers and receivers. With a similar end in view, surveys are being made of the State by counties to determine what crops predominate in each and the approximate quantities shipped during the season, together with a record of the places where shipments are made and the dates of shipping.

The last-named service has been undertaken for the consumer, as well as for the farmer, and should react favorably in preventing the abnormally high price periods that result from congestion and waste in the large shipping centers. The Bureau protects also the consumer by carrying out those provisions of the law demanding the inspection of licensed cold storage warehouses. Such plants are inspected at intervals of not less than two weeks and the handling of cold storage food by retail stores is inspected in so far as facilities permit. The Bureau gives advice to cities in the establishment of farmers' markets and assistance wherever investigation indicates that such a market can be operated successfully in the community.

The functions of the Bureau of Coöperative Associations are to assist communities in determining the advisability of organizing coöperative associations in their localities, to aid, where coöperative effort seems warranted, in organizing associations under Article 13 A of the Agricultural Law, with suitable articles of incorporation, by-laws and management, and to assist organizations once formed in installing simple but adequate systems of bookkeeping and accounting and to check up their annual reports and be certain they are proceeding along lines of sound and approved practice. The following are among the types of producers' organizations that have already been incorporated: onion-growers, potato-growers, hay-shippers, sheep-breeders, auto truck lines, associations for the operation of ditching machines and federations of associations of the growers of grapes, apples, peaches, quinces, and pears.

Consumers' coöperation is likewise encouraged. Field men of the Division respond to requests for assistance from organizations already formed, with the purpose of putting them on a firm financial basis. New consumers' organizations are formed and fostered by the Bureau in cases in which such procedure is deemed wise.

The Bureau of Weights and Measures has supervision over the inspection of retail weights and measures, of the sale of ice by weight, also of coal and of gasoline measures. New specifications and tolerances are established by the Bureau from time to time. Uniform city ordinances and advisable amendments to the weights and measures law are prepared as there is need for them. Lectures with lantern slides are frequently given before women's clubs and high-school audiences, showing the necessity for coöperation on the part of the consumer. Instructive exhibits are also prepared for use at fairs and expositions throughout the year as a part of the educational program of the Bureau.

The Bureau of Licenses has in charge the annual licensing and bonding of every person, firm, or corporation in the State receiving farm produce for sale on commission, or purchasing milk from farmers for shipping to a city or for manufacturing purposes.

Commission merchants are required by law to secure a license, for which $10 is paid and also to give a fidelity bond to the amount of $3000, which is filed with the Division to guarantee a correct report and the prompt payment to the farmer or shipper of all moneys due him. The milk dealer obtains a license at a cost of $10 and files a surety company bond or other security satisfactory to the Commissioner of Foods and Markets. The Bureau also has in charge the licensing of operators of public cold storage warehouses and issues such license only after inspection indicates that the warehouse is properly equipped and kept in a sanitary condition.

The Bureau of Food Products has supervision over the enforcement of the articles and sections of the Agricultural Law relating to adulterated vinegar, food and food products, honey, and maple sugar and sirup. Its agents are at work throughout the State collecting samples for chemical examination and for the determination of mislabeling. In close association with the Bureau of Food Products is the Bureau of Food Standardization which has in charge the chemical analysis of samples taken by the former and the formulating of definitions and standards for such food products as baking-powder, chocolate, cocoa, olive and table oils, peanut-butter, coffee, jams and jellies. Other coöperative activities on the part of the two bureaus include the following: investigation of the manufacture and sale of egg substitutes, inspection of bakeries and testing of samples of jams, jellies, shortenings, icings and fillings, with the idea of ascertaining the class of ingredients used and whether or not such ingredients are being sold in compliance with the food laws; giving attention to the sale of soft drinks; inspection of labeling of dried fruits and evaporated apple manufactures; inspection of the manufacture and sale of milk-powders and maple products.

THE CONSERVATION COMMISSION (See Fig. 35)

This department of the State government touches the individual farmer much less directly than does the Council of Farms and Markets, since it deals mostly with resources that are not the farmer's property. The touch of the Conservation Commission with natural resources through the supervision of the forests, the fish and game, and the inland waters has already been mentioned.

By exercising the inspection and supervision that are necessary to secure exemption or reduction in taxes on forest lands, it affects the farmer, and it aids him by supplying seedling trees at cost, for planting purposes. It may be expected that the farm woodlot will come to be looked on as a crop area as much as is the land in potatoes, hay, or apples, and there are indications that the net returns, though longer delayed, may be comparable with those crops.

The fish and game division aids the farmer in conserving the wild life on the farm and in the streams, that may be useful to him in protecting his crops, and it aids him in securing to himself the privileges to take game, animals, birds and fish on his own land if he so desires. By means of the fish hatcheries at thirteen points in the State, the streams may be kept stocked with desirable species. Four

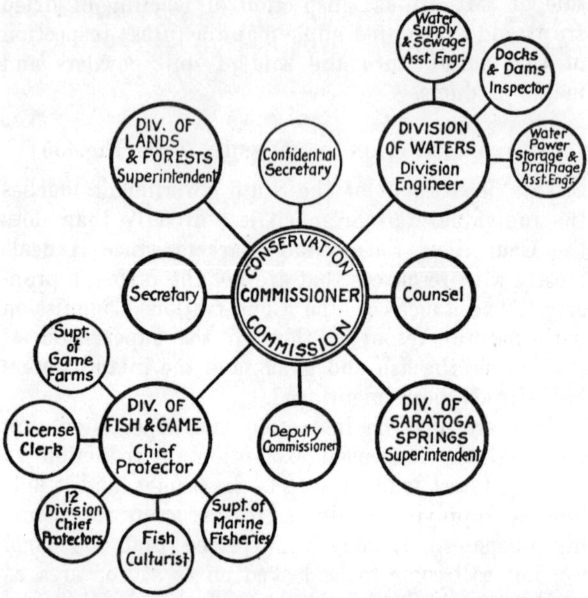

Fig. 35. Diagram showing the organization of the Conservation Commission into divisions and bureaus.

game farms are maintained to propagate game birds and animals and to some extent for purposes of investigation along those lines. Licenses for hunting, trapping and commercial fishing are issued by this division.

The division of inland waters has supervision over the streams, to insure the free flow of the water and to provide for its use for personal purposes and for water power. In order to protect the public no dam may be built except in the country and under ten feet in height, on streams having a flow of not to exceed three hundred cubic feet a second, without the approval of the Commission.

One feature of special interest to the farmer is the provision by which the services of the Conservation Commission may be invoked in draining and developing swamp land for agricultural purposes. Another and independent provision for the same purpose is under the agricultural drainage law. In these laws the conditions are stated under which wet lands and tracts of swamp may be drained by means of canals and ditches. By presenting a verified petition to the Conservation Commission requesting the investigation and improvement of any area of wet land, the machinery of that institution is brought to bear on the problem. Its engineers carry out the work and supervise the financial arrangements necessary. The cost of the improvement may be distributed over any number of years up to forty so that the financial burden in any one year is not heavy on the farmer. The fundamental authority for draining land for purely agricultural purposes and for acquiring such rights of way and for the distribution of the cost of such improvement over all the land benefited lies in Article I, Section 7 of the State Constitution as finally amended in 1919, after a long period of uncertainty under the earlier language of the Constitution.

Under the agricultural drainage law, the machinery of organizing and financing such a movement must be provided locally out of inexperienced persons and often with poorly trained help. The movement toward the drainage of a tract of land is instituted through the County or Supreme Court Judge of the district in which the improvement is located. The bonds authorized for meeting the cost of improvement under this law may be either township or district bonds and run for periods of five years until the total amount is paid.

It is important to note that the expense of draining and improving any wet or swamp land for agricultural purposes under either provision is not borne by the State but is assessed against the land benefited in proportion to the benefit received. Such improvement is a long-time investment. Under the Conservation Commission, the State does make some outright contribution in bearing the expense of the preliminary survey to determine whether the proposition is feasible.

These provisions are especially important to the farmer since, in addition to some two thousand square miles of swamp land in the State, there is several times as large an area now devoted to crops that needs partial or complete drainage to make it as profitable for crop production as possible. In many cases, the ditches, drains and dykes necessary for this purpose involve more than one owner and the law provides for the organization by means of which they can work together equably.

THE WEATHER SERVICE

The Weather Service in New York is provided in coöperation with the federal Weather Service. The State Director of the Weather Bureau is appointed and paid by the Federal Service, but the office facilities, clerical force, and maintenance fund are provided by the State. The office facilities are provided at the State College of Agriculture at Ithaca where complete weather observations are taken twice daily. Coöperating with this central office are voluntary and paid observers distributed over the State that report their observations to the central office, together with similar reports from other parts of the county that are the bases of the daily weather forecasts. The distribution of the weather forecasts from day to day are made in part through this central office. It is the business of the Weather Service to collect data showing the weather and climate, to distribute predictions on the weather twenty-four hours in advance, for the guidance of agriculture, shipping and other interests that may be concerned, and to tabulate and prepare for publication data bearing on the climate of all parts of the State.

CHAPTER X

THE EDUCATIONAL AND RESEARCH ORGANIZATION OF NEW YORK

THE educational system of New York is unique in its organization. Its administration and supervision centers in the University of the State of New York which touches every educational institution in the State from the smallest rural school to the great universities, libraries, and research museums. Without grounds, faculties or students in the sense in which those terms are commonly applied to the modern university, nevertheless it has all these in the institutions that come directly or indirectly under its supervision. In practical governmental parlance it is a Department of Education.

Through its various officers, the University of the State of New York encourages and expands the educational service of the State from the foundation of reading, writing and arithmetic to the cultural and vocational subjects that help each individual more readily to find his or her place in the work of the world, and at the same time secure something of that "bread and butter utility" of which each mature person must take account. It coördinates the various grades and phases of instruction and makes them an unified whole through which the pupil passes forward from grade to grade as he meets the successive tests. It unifies the methods and standards of instruction and provides for the preparation of teachers. It acts as the judicial authority in the arbitration of all disputed educational questions between pupils, parents or guardian and the local educational agencies. It does a certain amount of police work in insuring that every child receives the prescribed amount of instruction that the State considers essential to its welfare.

The educational system is removed from politics largely through its overhead administration by a non-partisan board of twelve regents, three more than the number of judicial districts, appointed from time to time by the Governor as their terms successively expire. This system of supervision began in 1784 and in 1904 absorbed the functions of a Department of Public Instruction.

Next to the Board of Regents is the President of the University and the Commissioner of Education, one chief officer in whom all the work of the University centers. The functions of the University, which may also for convenience be known as the Department of Education, are distributed in five groups: elementary education, secondary education, higher education, the State Library, and the State Museum. The first three are presided over, each by an assistant commissioner, the fourth by the State Librarian, and the fifth by the Director of the Museum. For administrative purposes, the Department is organized into thirteen divisions. The form of organization

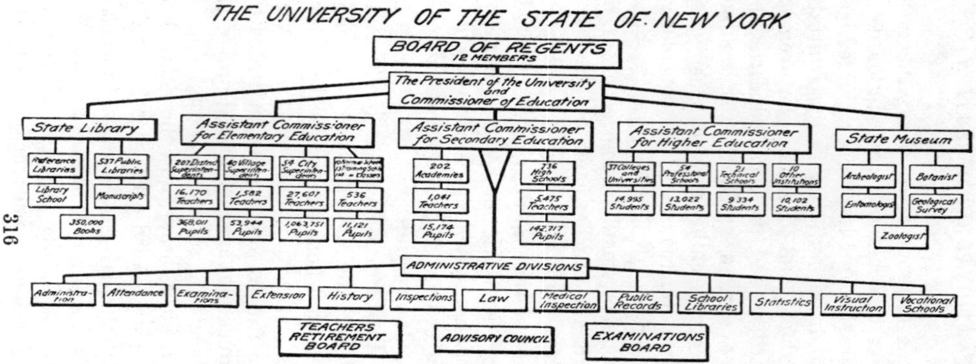

Fig. 36. Diagram showing the organization of the University of the State of New York and the numbers of units included in some of the main groups and divisions in 1915. This is said to be the largest single educational organization in the world and includes agricultural education as one of its professional and vocational types.

and the size and scope of the work are best understood from the diagram in Fig. 36.

In 1913, the State spent $83,869,254.97 on its educational system. For no other purpose do the people tax themselves so willingly. Of this vast sum, the State paid out of general funds $7,471,491.58. The remainder was derived from local taxation and from other sources. Over $50,000,000 was expended on elementary instruction in the more than twelve thousand districts (11600 in 1915 of which 8500 were one-room schools) and reached 368,011 rural pupils and 1,128,816 pupils in village and city schools. Secondary education in 202 academies and 736 high-schools reached 157,891 pupils at a cost of $13,569,115, about $34 to a pupil in elementary schools and $85 for pupils in secondary schools. Higher education in 122 universities, colleges and technical schools and other institutions cost $17,927,942 for 47,903 students or at the rate of $375 a student. The higher the grade of instruction, the higher is the cost for the individual.

The year 1912 was the centennial of the establishment of universal public elementary education in the State which is one of the cardinal principles at the root of the nation and the one to which more than any other the stability and permanency of democratic institutions is and must be due. It touches the whole people as nothing else does and lives with them through life. It is the basis of their inspiration, the tool in their work. One person in every nine is in the elementary schools; one in every eighty persons is in the secondary schools; and one in every two hundred and seventy persons is in institutions of higher learning. Omitting miscellaneous institutions other than colleges and universities, one in every three hundred and forty persons gains the opportunity for higher study. This is about three-tenths of one per cent of the population. Higher education is, therefore, an exclusive privilege to which all may aspire but few attain.

In 1910 the total illiteracy in the State was 5.5 per cent for all persons ten years of age and over. Considering all classes, it was 5.9 per cent for the urban population, and 3.9 for the rural. Among the native whites, the illiteracy is slightly greater in the country than in the city. For the United States as a whole, the illiteracy was 7.7 per cent which is 2.2 per cent greater than in New York in spite of the large influx of foreign population on whom the ability to read and write is imposed as a condition for admission by the federal immigration laws.

The organization and administration of the city and village schools for a long time very much surpassed that in the rural communities. Always there has been coöperation between the local forces and the State system in conducting educational work. The educational system has been likened to a small state. Each local group is presided over by trustees and boards of education in the cities, and by school directors in the country districts. The city public schools are directed by the superintendent. In the last few years, a similar measure of supervision has been extended to the rural schools through the system of district superintendents. These officials, who must be qualified and experienced teachers according to the standards of the Commissioner of Education, must also be qualified to give instruction in agriculture. Under the district superintendents is placed a convenient group of townships and all the schools in this unit come under his or her immediate supervision. The educational system has been the leveler of franchise and suffrage standards, for men and women alike have access to all its positions from the highest to the lowest, and for many years before general suffrage was extended, women, with certain restrictions, were permitted to vote on all educational matters. School books may be furnished to the pupils at public expense if so elected by the local board of education and this practice prevails in many places. The centralized supervision and close grading of the instruction makes possible the Regents system of uniform examinations and rating in all elementary and secondary schools throughout the State as a result of which pupils graduating from the public school system including the high-schools receives a state certificate of proficiency.

In recent years, special attention has been given to vocational work in the grades and secondary schools. Like other forms of education, this has diffused down from the institutions of higher learning. Vocational training takes three forms according to the character and status of the pupil; industrial training particularly for the city boy; agriculture for the country boy, and home economics for the girls, along with such agricultural or industrial instruction as they may elect to take with the boys. In the grades this instruction assumes the simplest forms and is designed to put the pupil in sympathetic touch with his environment. Nature-study is a popularized term for the introduction of nature subjects to the pupil. The fundamental aim always is or should be to lead out the mind of the pupil, to induce it to react in thought on the things with which it comes in contact, and not only to catalogue and classify these but to see and understand existing relations or establish new relations. This is the business of life and the larger the power of observation and the broader the capacity for arrangement and coördination, the better educated is the person.

In the country in particular, but in the cities as well, it is increasingly recognized that the first materials of education are the natural objects about the pupil with which every person has constant contact, and which provide practically every element of training through their study. New York has perhaps been a leader in carrying the nature-study point of view to the grade school pupil, especially in the country, and hundreds and even thousands of boys and girls and men and women in the State of New York carry with them through life the zest and enthusiasm and outlook that had its inception in the pioneer work of Uncle John Spencer, of Westfield, and of Cornell University, who by his introduction in the simplest way and with a most sympathetic touch made the garden more inviting, the corn rows less long and the chores of the kitchen and stable more acceptable. This will solve some of the difficult problems of the rural schools, although introducing others that are objectionable to some parents.

In addition to the very small amount of nature-study instruction given by the teacher, each of the 207 district superintendents devotes eleven months of the year to his work, and supervises home agricultural projects and exhibits for the aid and encouragement of the pupil, the teacher and the parent during the summer months. It is now provided that any township may centralize its school system at one point with a corresponding increase in teachers and equipment, with provision for the transportation of pupils from the remote sections.

In secondary education, the vocational work begins to take definite form in courses. The industrial and domestic science subjects in the cities have become an established part of the instruction and have been extended into night schools for the older classes and the continuation schools for those who desire to maintain a touch with the school room after they have been forced or perhaps led on into commercial pursuits. Thus the break between school instruction and life work is closed and the latter follows in the natural evolution from the former.

A tremendous impetus has been given to vocational training in the secondary and grade schools, not only in New York but throughout the United States, by the passage in 1917 of the Smith-Hughes

Act by which the federal government makes a large money contribution to each State based on its population. This contribution is duplicated dollar for dollar by state appropriation for the promotion of vocational instruction and for the training of teachers for that purpose. The quota of federal money received by New York under the provisions of this act amounted in 1920 to $360,663, and will increase until 1925.

Agriculture is taking its place on a par with industrial training in secondary institutions, as an integral part of the educational system. Courses of instruction above the grades are always elective to the pupil and there is no purpose to force the individual into any particular groove but to afford him the freest and most natural opportunity to choose his life work and to take account of the probabilities based on situation and sex.

Agriculture of a secondary character is being more clearly recognized to belong in the vocational group and to be placed properly in the high-schools. Vocational work is only a part of a course and must be prepared for and backed up by cultural and training work. This backing the high-school already affords. Not every high-school should necessarily give instruction in agriculture but enough schools should be selected to meet the needs of every part of the State at points nearest home and best situated to give such work.

Encouraged by State aid in the payment of teachers of agriculture and home economics, sixty-nine agricultural high-schools have introduced such courses up to 1919. The State pays two-thirds of the salary of the first agricultural teacher up to a maximum of $1000, and an added amount for an assistant or a teacher of home economics.

Higher education in New York is fostered by thirty-seven colleges and universities, fifty-four professional schools, twenty-one technical schools and ten other institutions. Several of these are grouped in affiliated institutions, such as Columbia University, Syracuse University, and Cornell University. There are nine institutions for the exclusive education of women.

The State Library, together with the State Museum and the administrative offices of the University of the State of New York, are housed in Albany in the State Educational Building, a splendid structure completed in 1912. It is a parthenon of education. The name of the State Library sufficiently explains its function.

The State Museum is not so clearly defined by its name, since in addition to the preservation of material and data, it has five research divisions, botany, zoölogy, entomology, geology, and archeology, for original investigations, from which issue reports and bulletins.

At the head of higher education in agriculture is the State College of Agriculture at Cornell University, at Ithaca at the head of Cayuga Lake in Tompkins County. The beginning and development of Cornell University is especially interesting because of the affiliation it has always had with agriculture, and the leadership it has exercised in directing scholarship to the study of the sciences and arts, as well as to the humanities.

The federal Land Grant to the states under the Morrill Act of 1862, for the establishment of agricultural and mechanical colleges, was in New York utilized in a special way. This beneficent act found in the New York Assembly of 1865, to which it came for acceptance and disposition of the grant, two men of broad and high ideals who supplemented each other and admirably coördinated the State with the provisions of this act. Andrew D. White, a graduate of Yale University, held ideals for the vitalizing of educational methods and the broadening of its facilities to take in the sciences and the arts. Ezra Cornell, a successful man of business and possessed of interest in the development of agriculture, was willing to put his personal fortune with the federal grant and to manage the latter for the benefit of the State in the founding not only of a College of Agriculture and Mechanic Arts but "A university where any person may find instruction in any study" and of which agriculture and mechanic arts are a part. These public and private funds were pooled to establish a new ideal in higher institutions of learning. Whether consciously or unconsciously, this new institution conformed to a principle that has since come forward more clearly, that vocational work attains its best strength when coupled with and backed up by the humanities and the sciences that underly much of the handwork of such professions. The process of evolution throughout the country has been toward the University rather than separate colleges of mechanic arts and in recognition of this principle the universities of other states have in many cases grown out of the Land Grant Act of 1862.

The Land Grant Act was a federal provision to which New York for a third of a century made no substantial contribution. Agriculture during that time was a coördinate department in Cornell University. Then in 1904 the need for larger support of the agricultural work to meet the growing demands of the rural population, led to the adoption of this department by the State and its transformation into the New York State College of Agriculture, still coordinate with the other colleges in the University, and with the continuation of the support it had received from the federal Land Grant and the private endowment of Mr. Cornell through Cornell University.

From that time the State has made increasing annual appropriations for the maintenance and upbuilding of the College that have enabled it to expand its equipment and facilities more nearly to meet the needs. The College of Agriculture is an integral part of the University and is administered by its board of trustees, which was enlarged at the time of the creation of the State College of Agriculture, by the provision for the ex-officio membership on the board, of the governor, the lieutenant-governor, the speaker of the assembly, the commissioner of education, the president of the state agricultural society, the commissioner of agriculture, and a representative of the State Grange, together with five persons appointed at large by the governor. The remaining twenty-seven trustees are chosen by the alumni or elected by the trustees in office, with two exceptions. Thus, the governing board of the University, including the State College of Agriculture, is divided between persons who represent the public interests of the State, and those who stand for the private interests of the corporation and the body of alumni, an arrangement that would seem to secure the advantages of both types of direction with a minimum of the disadvantages of either acting separately. A special executive committee or council of the Board acts as the intermediary in the administration of the College of Agriculture and works in close coöperation with the Director of the College.

The objects of the College of Agriculture as stated in its charter are defined to be "to improve the agricultural methods of the State, to develop the agricultural resources of the State in the production of crops of all kinds, in the rearing and breeding of live-stock, in the manufacture of dairy and other products, in determining better methods of handling and marketing such products and in other ways; and to increase the intelligence and elevate the standards of living in the rural districts."

Under the charter the institution stands for the broadest possible development of the agricultural interests of the State in people and materials. To carry these functions, the College is organized into twenty-three departments of study, besides the work that is secured to the students from other colleges in the University. These departments embrace botany, farm crops, forests, fruits, flowers, vegetables, animal, dairy and poultry husbandry, insects and plant diseases, soils, agricultural chemistry, farm mechanics, rural art, farm management and rural economics, rural sociology and home economics, besides other related departments.

A farm of 1128 acres, with barns and live-stock, supplement the twelve buildings, having an aggregate cost of about $1,000,000 and house these departments with their equipment and facilities. Two hundred and ninety-three regular courses of instruction in agriculture and thirty in home economics are offered for the election of the student body. These departments and courses of study illustrate the wide diversity of problems with which rural people deal and indicate the complexity of the relations and process with which the farmer may be confronted on a single farm. It is, therefore, essential that the individual who would take part in agriculture be trained for the work. A prime reason why agricultural training and organization have been so long in developing is this very variety, which must wait on the development of the related sciences and arts to lead the way to the understanding of the facts and processes of the farm. This same variety and complexity indicate the limitations of the individual farmer, working alone on his own farm, and suggest the desirability of such a reorganization of the farming business as shall permit specialization of function in the individuals who take part in agriculture along the lines of their personal training and qualifications. Like a great machine made up of many parts and each with its particular work to perform, so agriculture would seem to be a business best carried on in such sized units as will concentrate volume, apply special skill and insure continuity of work. The educational development of the subject foreshadows the economic and social development. The opportunity for real rural community building to provide adequately for all these needs and limitations is large and everywhere at hand.

Agricultural colleges have taken the lead in the appreciation of the importance of the personal well-being of the workers because of the intimate relation that has always existed between the rural business of agriculture and the intellectual, social and moral life of the people engaged in it. A farm is usually a home as well as a business.

In addition to the University course of instruction based on the usual high-school preparation and leading to the degree of bachelor of science, the State College of Agriculture has found it desirable to offer shorter courses of study based on less preparation. These have been forced by the lack of facilities for such instruction in the secondary schools and have been the forerunners of such instruction. Students are admitted for special courses of study and for short or winter-courses of a few weeks. Such students are now required to have reached the degree of maturity indicated by an age of twenty-one years, and

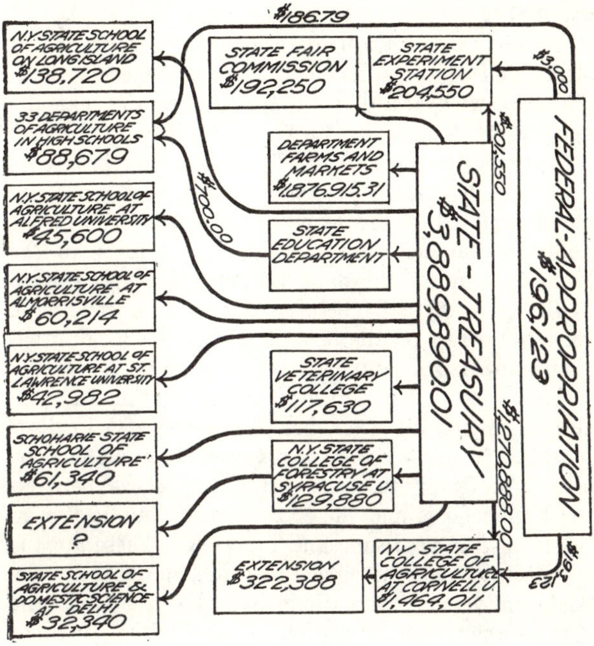

Fig. 37. Diagram showing the amounts and distribution of federal and state expenditures for agriculture in New York State.

must have had farm experience. As the facilities for training in agriculture in the secondary schools develop, the need and the call for such courses in the

State College may be expected to decrease. The instruction given is usually above the standard of the secondary schools because of the maturity and wider experience of the students.

It is notable that home economics as a branch of study and as a professional art has generally been affiliated with training in agriculture and has developed hand in hand with the recognition of the natural humanitarian aspects of the rural work. It is, of course, obvious that home economics is not more closely related to the home of the farm than to the home of the merchant or the manufacturer in the city or to that of the mine worker. The recognition of the needs of the home-keeper has been a little more easy and direct on the farm and thus the rural situation is making a contribution to the urban.

The demand for intermediate instruction in agriculture and home economics has given rise in New York, as in other states, to a series of hybrid schools, already referred to, that stand in a sense between the high-schools and the State College but nearer to the farmer in operation and function. These are the six State schools of agriculture distributed over the State in various affiliations and distinct from the regular system of education. The first of these to be created was the State School of Agriculture affiliated with St. Lawrence University, and administered by it, at Canton in St. Lawrence County. Persons prepared to enter high-school are received here as at all other state schools and given a course of training of three years while students having some high-school training may finish the course in two years. In many ways, the instruction resembles that given in the college as well as that in the high-schools. The work is done in laboratories and on the farm. The School of Agriculture of Alfred University at Alfred in Allegany County is very similar in organization and operation.

The other four State schools differ from those mentioned in not being affiliated with any other educational work, and are special vocational schools. In the order of their creation, they are located at Morrisville in Madison County, Cobleskill in Schoharie County, Farmingdale on Long Island, and Delhi in Delaware County. All have farm lands and stock, in addition to buildings and laboratories for instruction. All are authorized to carry on tests and do educational work beyond their walls in the adjacent territory. With the exception of the School at Canton, there has been some provision in the organization of the board of control for the coördination of the work of the school with the other institutions in the State giving agricultural instruction, by the ex-officio appointment of the Dean of the State College of Agriculture and the Commissioner of Agriculture. In 1918 all these schools were put under the immediate supervision of the State Department of Education as a part of the secondary school system of the State. If such special schools are to be established to deal with regional problems, they should be located with due regard to the particular agricultural interests they ought to serve, with the view to keeping the number to the lowest practical terms because of the relatively large expenditure required for their equipment and maintenance.

The teaching of domestic science is an integral part of the work of all these schools. No degrees are conferred by them. They are entirely supported at State expense and their work is carried on by a director and the faculty assembled by him, except that in the schools at Canton and Alfred the president of the University is given some prerogatives.

A State College of Forestry was established at Syracuse University in 1911, to give instruction in forestry and forest management and allied branches. It is administered by a board of control made up, as ex-officio members, of the chairman of the Conservation Commission, the Commissioner of Education and the Chancellor of Syracuse University, together with three persons appointed by the Governor and six appointed by the trustees of Syracuse University. It is, therefore, only indirectly affiliated with the system of agricultural education in the State. It offers a five-year professional course leading to a degree of Master in Forestry, and a two- or three-year ranger course, the latter on the one thousand acres of forest land controlled at Wanakena on Upper Saranac Lake in the Adirondacks.

Other agencies of college training in agriculture are a department in Columbia University in New York City and the David Slocum College of Agriculture in Syracuse University. They are designed to offer agriculture as a part of the curriculum, primarily as a branch of applied science along with other subjects of university instruction. Both schools are privately supported and have farm lands and equipment, but the latter is not affiliated with the College of Forestry at Syracuse. The work in these colleges leads to a recognized degree at the end of the prescribed course of study. Their tuition is that usually charged by such private institutions.

Two other institutions not strictly agricultural in their work but closely affiliated with it in their practical aspects are the state schools of veterinary medicine. The first of these is the New York State Veterinary College at Cornell University established in 1894 and located in Ithaca. It is one of the coordinate colleges of that institution on a footing with the College of Agriculture. Its objects are stated in its organic law to be "to conduct investigations into the nature, prevention and cure of all diseases of animals, including such as are communicable to man and such as cause epizootics among live-stock; to investigate the economic questions, which will contribute to the more profitable breeding, rearing and utilization of animals; to produce reliable standard preparations of toxins, antitoxins and other products to be used in the diagnosis, prevention and cure of diseases and in the conduct of sanitary work by approved modern methods; and to give instruction in the normal structure and function of the animal body, in the pathology, prevention and treatment of animal diseases, and in all matters pertaining to sanitary science as applied to live-stock and correlative to the human family."

Veterinary medicine bears the same relation to animal husbandry as do plant pathology and economic entomology to crop production, the latter being a part of the College of Agriculture. By the terms of the laws under which these several Colleges of Agriculture and Veterinary Medicine are created, there may be some over-lapping in the functions they are authorized to perform. In their practical operations, they are supplementary institutions.

In 1913, the State adopted the School of Veterinary Medicine of the City of New York as the State Veterinary College for the eastern part of the State under the name of the New York-American Veterinary College, and its objects are stated to be substantially the same as those of the Veterinary College at Cornell University. Thus far, it has received very little financial aid from the State.

The course at Cornell is four years in length and leads to a degree and admits such graduates to the practice of veterinary medicine in the State. At the latter institution the course is three years. In addition to the laboratories, clinic and instructional facilities possessed by both institutions, the College at Cornell has a farm of one hundred and forty acres for the isolation and study of contagious diseases. The latter College also prepares and furnishes at cost all the serums and antitoxins used in the treatment of animal diseases. A feature of the work of the institution is instruction and laboratory practice in scientific horse-shoeing.

In all cases in which State appropriations are made for agricultural instruction, whether to independent institutions or to those affiliated with other educational organizations, it is required that the land upon which state buildings are placed, together with the necessary right of approach to them, shall become the property of the State.

EXPERIMENT STATIONS

Schools presuppose subject-matter to teach. As a matter of fact, the systematized methods of thinking essential to good teaching have revealed the gaps in the available information and have sent the aggressive teacher into the investigational field to push out the borders of knowledge. The beginnings in agricultural education were especially confronted with the lack of teachable material which was partially responsible for the first quarter century of slow progress made and the small recognition received. The Hatch Act passed by the federal government in 1888 provided for the establishment of experiment and research stations in the several states to work out the problems of the farmer and provide material for instruction in the agricultural colleges. These, too, groped their way along for a period until the field was organized on the new basis of function and science instead of on the basis of farm practice, as was at first attempted. Agriculture is applied science and its accomplishments constitute the art of agriculture. New York anticipated by eight years the Federal Experiment Station movement by establishing a station in 1882 on its own initiative, and located at Geneva; and the staff in agriculture at Cornell organized the Cornell University Agricultural Experiment Station in 1879 and began the publication of bulletins. This Geneva station, with its 130 acres of farm land, with laboratories, stables, stock and an excellent staff of workers, has been a leader in agricultural investigation in the country. It has conducted a wide variety of studies. Its chief departments of investigation have been horticulture, agricultural chemistry, with particular reference to animal nutrition, botany, pathology, entomology, farm crops and dairy industry. Over five hundred bulletins had been issued up to 1919 besides circulars and special reports. The general control of the institution is vested in a board of nine members appointed by the Governor which acts through a director who is responsible for the policy and administration of the institution. In addition to the studies at the station, investigations are made in different parts of the State according to the nature of the problem in hand.

This Station is supported entirely by state appropriations, with the exception of $1500 that is received annually from the federal government under the Hatch Act that secures to the station the use of the franking privilege in mailing its publications.

New York is one of the few states that has two experiment stations. The federal appropriations for experimental and research purposes have been assigned by the legislature to the State College of Agriculture, except the sum mentioned above, that grants the franking privilege to the State Station at Geneva. This is called the Cornell Agricultural Experiment Station and the investigations are conducted by the staff of the College of Agriculture at Ithaca. The funds appropriated under the Hatch Act of 1888, by which the sum of $15,000 is available from the federal government, are used by the several departments according to the designation of the Director of the Station, who is also Dean of the College, in the conduct of experiments and investigations, and is the basis of the bulletins issued from time to time. In 1906 this appropriation was supplemented by a second federal enactment appropriating funds to the several states known as the Adams Act and designed to stimulate the deeper and more fundamental lines of investigation and research relating to agriculture, in distinction from the experimental and local work that might be carried under the Hatch Act. This latter act recognized the need for fundamental investigations into the facts and principles underlying agricultural practice of general rather than local application. The sum assigned to the Cornell Station is used for investigations in three lines: soil technology, plant-breeding, and economic entomology. Certain members of the staff in these departments give practically their entire time to such investigations.

Under the stimulus of these two funds, in addition to the encouragement given to investigation by the regular teaching staff of the College of Agriculture, somewhat over four hundred bulletins have been issued up to 1919 that cover practically every phase of the field of agriculture. Recently a new series of publications has been instituted known as the Memoirs that is devoted entirely to publications of a research nature and contains much technical detail that is useful to investigators and special students in the same field, but it is not adapted to reading by the general public as is the design of the regular station bulletins.

To carry the data accumulated by the experiment stations and by the general agricultural investigations to the farmer on the land, several agencies have been called into existence. The earliest of these, outside of the agricultural press which has long been a potent factor, is the agricultural bulletin. These report the results of special studies from time to time. However, special bulletins, while of large value, especially to the person in general touch with the field of study, do not altogether meet the need of the individual who has not had special agricultural training. He needs something to take the place of a well rounded course of study. Often he has not been accustomed to find such information in books that are now available dealing with nearly every phase of agriculture. He, therefore, has use for general bulletins that discuss and explain in an untechnical way the several fields of his work. These extend to the farmer the information given in schools and colleges of agriculture and by a natural process of evolution are called

Extension bulletins. They issue from the College of Agriculture and, in the form of supplements to the regular bulletins, from the Geneva Station. One of the later forms of these bulletins is the Reading-Course bulletin for the farmer or for the farm home, issued by the State College of Agriculture. They are a series of lessons prepared by the different departments that most closely touch farm work, and in the successive issues that are published from time to time carry forward a somewhat consecutive line of study. There are two divisions in these bulletins, one for the farm and the other for the farm home. The other schools and colleges in the State are not authorized to put out any general series of bulletins but may issue leaflets dealing with their particular fields of work.

EXTENSION WORK

In recent years a general extension movement has been projected, not only in New York but throughout the country, to reach the farmer and his wife who may not be in position to go to a college or other agricultural school. It had its origin in a variety of disconnected movements, one of the earliest of these being the Farmers' Institute under which plan one or two day schools are conducted by a staff of one or more persons that discuss and explain lines of work that are applicable to the region. As was indicated in an earlier chapter, the Farmers' Institute formerly issued from the State Department of Farms and Markets but by act of the Legislature in 1918 it were transferred to the State College of Agriculture at Ithaca and closely coördinated with the system of extension teaching in operation there. The Farmers' Institute reaches into hundreds of communities and carries the gospel and the inspiration of better farming. These meetings reach about a hundred thousand persons annually during the institute season from November to April. With the development of other forms of rural teaching, the Institute has come to be looked on more particularly as the pioneer movement in any given region to reach persons who would not otherwise come in touch with the instruction for agricultural betterment. But even in the older communities, the Institute still fulfills some of the functions of an annual get-together of the agricultural interests for the discussion of local problems.

The speakers on the staff of the Farmers' Institutes are successful farmers, teachers and investigators from the colleges, schools and stations relating to agriculture, and other persons who may be called in because of their special knowledge and training. The salary of the speakers and their expenses are borne by the State. The remaining expenses are assumed by the community.

Miscellaneous lecture work is carried by the staff of nearly all the agricultural colleges, schools and experiment stations at such points in the State as may request aid. In this way, thousands of persons are given an enlarged vision concerning some phase of agriculture. Some of these lectures, which are arranged by churches, communities, associations, commercial boards, schools and other working groups, are made up of a series to continue at intervals for days or weeks or even for an entire season.

It is now the stated policy in most of this lecture and demonstration work from state institutions, including the State Department of Agriculture, that the community requesting such aid coöperate by meeting a part of the expenses of the speaker, not including salary. When an individual or a community contributes to such expenses for its benefit, better use is generally made of the service rendered and many perfunctory calls for mere entertainment are eliminated.

A still further development of the extension idea is in the farm demonstration schools. These aim to reach the more mature persons in a community and to present to them a somewhat comprehensive course of study, usually of one week's duration. These, also, are carried by the State College of Agriculture. During the winter months these demonstration schools have been held to the extent of forty to fifty a season and treat three to five subjects. The instruction is consecutive and constructive and aims to build up a fairly sound general understanding of the subject in hand, and to point out its application to the conditions in that region.

Another educational movement to reach the man on the land is the Farmers' Week held by the State College of Agriculture and by the older established schools of agriculture. These one-week meetings at the institutions, for farmers and their wives, aim not only to give information which is imparted in lectures, demonstrations and field and stable gatherings, but to familiarize the farmer with the equipment, methods and work of the institution to the end that there may be a better mutual understanding and greater unity of effort. This type of activity grew up at the College from a very small beginning in the annual conference of farmers interested in experimenting on their own farms and has now become a large enterprise that annually brings several thousand persons to the College. This institution has now been adopted by the schools of agriculture and the colleges of other states.

The last and perhaps the most fundamental and far-reaching step that has been taken to coördinate the farmers of a community and to give them the benefit of agricultural investigation and experience in the light of their local conditions, is the Farm Bureau enterprise. All other movements have thus far largely lacked close contact and understanding between the rural people and the state institutions for their aid. The problems are too diverse, the distances too great, the facilities of the institution which are largely consumed in other directions are inadequate to give the local contact with and understanding of the problems and queries presented by the farm that are essential to effective aid. There was need of an agency to stand between the farmer and the various state and federal sources of information and aid. Such an agency should be headquartered locally and should be familiar with local conditions of the farming region. It should bring the current information on rural questions to the individual either directly out of the experience and training of the local leader or through coöperation with the proper department of the state or federal institution. Some sort of an agricultural clearing-house arrangement was essential and this the Farm Bureau provides. In New York State the movement developed out of an effort to secure this local coordination in Broome County in March, 1911, where a college-trained man was placed by the combined effort and support of the railroads, the Chamber of Commerce of Binghamton, the State and Federal Departments of Agriculture and the State College of Agriculture. The duties of this agricultural agent, who had an office in the Chamber of Commerce rooms of the chief city in the region, was to become acquainted with the farmers and their conditions, to consider their queries so far as he could, to call in outside help when necessary and in general to promote the coördination and development of all the rural agencies in an area which was roughly marked off by a boundary of forty miles around the city of Binghamton.

Other communities seeking the same sort of help took up the idea and worked it out in a slightly different way, particularly with reference to its financial support, and a definite policy began to be formulated. In Jefferson County the whole population first gave support to the movement through a contribution from the County Board of Supervisors, together with the other sources of support. It became a real bureau directed by a representative board. A little later followed the organization of the farmers of the county into a farm bureau association, with annual contributions from each member, and with advisory members or committee-men in each community center in the county. When all these local and state forces were finally tied together by a state leader and advisor, the present Farm and Home Bureau system was completed. The most recent development has been the home bureau for women and with a woman demonstration agent coördinate with the organization for men.

The organization of a Farm and Home Bureau association, with one or more trained men and women in charge, is a matter for local initiative. The advice and help of the state leaders may be sought, but the responsibility for its inception and support rests with the people of the county which is now made the unit. The county leaders are chosen from an approved list of candidates supplied by the state leader, by the county organization which provides the physical equipment of the Bureau. The bureau agent is now made the medium for the aid furnished by the several governmental institutions — the experiment stations, colleges of agriculture, the state departments of agriculture, and the federal Department of Agriculture. All the bureaus are drawn together and the experience of one made available to others through the state leader's office which is established at the State College of Agriculture as a regular part of the staff of that institution, and, therefore, as an official part of the State Extension service. On the farmers' side the several county farm bureau associations of the State are now drawn together in a state federation of farm bureaus and with similar organizations in other States is consolidated into the American Farm Bureau Federation.

In financial support, a definite policy of coöperation has been developed. The funds from the federal government under the Lever Act and from the State are joined with the contribution by the board of supervisors, the membership fees of farmers in their association, and other local contributions. As a means of carrying information and inspiration to rural people, of aiding in their organization for more effective business, and in meeting other needs, the Farm Bureau movement has been a big impetus. It promises much more and seems likely to be an effective coördinating and expanding force in rural life. It will, of course, take time for such an institution fully to find its place, and in the meantime expectations may occasionally be unduly inflated. But the movement must ultimately take a firm place in rural welfare.

Mention should be made of the funds now put into the several states by the federal government under the provisions of the Smith-Lever Act of 1913. The primary purpose of this act was to carry the information already secured to the man and woman on the land in a more effective way. The promotion of home economics is a special object of the act. Appropriations under this act began in 1914 and after 1922 will amount to $195,000 annually for New York, based on the rural population. This distribution is subject to revision on the basis of each federal census. To be available to any state, the federal appropriation must be accepted by the state legislature and assigned to the State College of Agriculture, and it must be matched by an equal annual appropriation on the part of the State for the same purposes and under the same conditions. Thus, about $9,000,000 will be available to the several states in and after the year 1923 for the purpose for which the appropriation is made. It is to be used in educational and demonstration work away from the state institutions. It is purely an extension movement and may be applied to any of the lines of extension work mentioned above. The fund is administered in New York by the State College of Agriculture, subject to the general supervision of the United States Department of Agriculture.

New York was probably the first of the states to engage in extension work in agriculture under the authorization of a special grant for the purpose by the legislature. The first appropriation, for such work, under the so-called Nixon Bill (named for the late S. F. Nixon, Speaker of the Assembly), was in 1894, now more than a quarter century ago.

OTHER EDUCATIONAL ORGANIZATIONS

New York State has a large number of organizations touching the agricultural interest. A summary

Plate VIII. View in the Mohawk Valley region. General farming and dairying.

of these was made in 1911 in a bulletin of the State Department of Agriculture, at which time there were 1096 state and local organizations of one sort or another relating to agriculture. Forty-one were organized to reach the entire State. Of the remainder of 1055, ten were state educational and administrative institutions applied to smaller divisions. At the time mentioned, 1911, these organizations represented the following interests: 815 were Granges, 94 fair societies, 36 Patrons of Industry, 24 poultry associations, 23 farmers clubs, 15 dairy associations, 13 each of cow-testing associations, and agricultural and horticultural societies, 11 each of fruit and stock-breeding associations, 10 governmental experimental and educational institutions, 9 coöperative associations, 8 beekeepers associations, 7 plant-breeders and crop associations, 4 driving clubs, and 1 each of drainage clubs, hop-growers associations, and housekeepers' conferences. Since that date, there has undoubtedly been considerable shift. The total number of organizations has materially increased and this would be especially marked in the coöperative associations, cow-testing, farm bureau and other special welfare organizations and in the local granges. If the members of these associations lived up to the ideals stated in their charters, the rural situation would be a paradise, providing the conflict on their borders did not bring pandemonium.

The Grange came into existence at a trying period in the development of the country and established its first local organization at Fredonia in this State in

1867. It aims to inculcate the appreciation and love of country life, to instruct in the science and art of agriculture, to enlarge the intellectual horizon in the home and on the farm, to promote economic efficiency and social welfare and to protect and minister to its members. The county, state and national system of organization help to maintain and guide the local branch, as well as to give force to its wishes. The primary efforts of the Grange have been directed to social, educational and protective movements. Economic organization has thus far received only casual attention or in some cases movements instituted, such as Grange stores, have worked themselves out of existence through unwise organization or poor leadership.

In 1918 the State Grange Exchange was organized with the approval of the State Grange, as a general business corporation directed by officers and directors of the Grange. As noted above this has been merged with the Grange-League-Federation Exchange which is a State-wide movement.

Much influence has been exerted on public questions affecting the farmer and generally in a constructive way. The Grange was one of the early national equal suffrage organizations in the country, and has been an effective educational agency in that respect. As its members gain a broader outlook and see more clearly the various functions they have to perform, and use the local Grange units in those directions, it will fulfill a much larger service. The fact that it has a secret ritual keeps many good persons out of membership for personal reasons, and limits its usefulness, and on the other hand it is a source of attraction to others and by that contact a means of help to many. The fraternal spirit, as expressed in such institution, is in a sense the spirit of exclusiveness or cast, or aristocracy. It has served as an effective lever in welding individual sentiment, but it must grow into a broader spirit of inclusiveness.

The Patrons of Industry is a younger (1887) institution than the Grange and is organized in a like manner and with similar objects. Its membership is very much smaller than the Grange. Fire protection is a prominent feature in its program. The organization has generally developed where the Grange did not exist.

There are twenty-nine state organizations dealing with particular divisions of plant or animal production. These include the various stock-breeders' associations for the different types of live-stock, dairy, poultry and beekeepers' associations, and crop associations prominent among which are the fruit-growers and the vegetable-growers. Antecedent to all these is the State Agricultural Society, which was for many years the chief clearing-house for agricultural discussion and progress. The breaking up of the State organization into working units allied with special interests is a symptom of progress, preliminary to a new general alignment of rural interests. Recently the society has been revived in affiliation with the Department of Farms and Markets as a sort of general agricultural forum, or central clearing-house for all the other agricultural societies in the State. In recent years its reports have been published as bulletins of the Department of Agriculture.

The effort to represent agriculture in a State board or society appears first to have taken form in the "Society for the Promotion of Agriculture, Arts and Manufactures," founded in 1791 and chartered by the legislature in 1793. The Society was to collect information on agriculture from the counties for publication. The charter having expired in 1804, the legislature chartered "The Society for the Promotion of Useful Arts in the State of New York." In 1818, Governor De Witt Clinton advised the organization of a board to represent agriculture and closely allied interests and to further them by means of education. In 1820 the Board of Agriculture was founded. It was to work with the county societies, to distribute seeds, award premiums and otherwise "to promote the agricultural and manufacturing interests of this State." It expired by limitation in 1825, and in 1832 the State Agricultural Society was founded. The idea of county exhibitions and reports was continued through all these changes. This society has had an honorable and influential career, and many educational and similar movements have come out of it. In later years it has found a less important place to occupy because, for one thing, the State Department of Agriculture (now the Department of Farms and Markets) assumed the relations with the counties as expressed in the fairs, and the State Fair is no longer part of its work.

The local and county fairs are in a mixed condition. Sometimes they have degenerated into a questionable sort of amusement, dominated by horse-racing. The lack of affiliation with the year-round agricultural interests in the community to give the work permanency, and the failure of the premium lists to take hold of the life interests and the economic units of the farm, have been largely responsible for such degeneration. It is interesting to note that in the early days prizes were offered for entire crops, for field and farm yields and for herds, and not merely for specimens of product. As a part of the general community organization and as the annual testing ground for these, the fair might hope to get back to a position of relative leadership. Some censorship in the multiplication of such associations might well be exercised by the State Division of Agriculture to secure the most effective units.

The State Fair at Syracuse, under State financial support, is building up a magnificent equipment in grounds and buildings. This is used only one week in the year, and is an unprofitable bill of expense the remaining fifty-one weeks. Some method should be found at least partially to use this equipment for other exhibits, gatherings and agricultural sales of all sorts. Under the supervision of the State Department of Farms and Markets, and in coöperation with the State institutions, it might be made the central exchange and perpetual show ground of the agricultural interests and the center of a larger educational service.

Among animal interests, the Breeders' Association and the Dairymen's Association are the largest, with 400 members each. In horticultural interests the New York State Horticultural Society — with 1700 members — is the largest not only in the State but also in the country. They meet once or twice a year.

A further feature of rural organization is buying and selling associations, the formation of which has been much stimulated in the last five years and appears to be making marked progress.

In addition, there is a very large number of smaller local associations, such as reading-clubs, social clubs, canning clubs and other welfare organizations, all of which promote the spirit of mutual understanding and are the stepping stones to broader work. Each individual must choose the organizations and movements that most nearly represent his interests and ideals and that most effectively give force to them, and it may not be practicable for him to touch all of the organizations in his community. Not to be affiliated with any of these is an indication of either narrowness or selfishness, and in both cases of underefficiency.

New York, throughout its history, has been well churched. It is a heritage that it derives from the New England stock, as well as from colonists from other sources, as indicated in the chapter on history. The rural villages, which are usually the seat of the church, have one to several churches representing different denominations, and in the aggregate practically all of the denominations extant in the country.

justed many complaints of shippers of agricultural produce, although, of course, it cannot make regulations or provide facilities that are not reasonably warranted by the freight to be handled.

Less than a quarter of the products of the farm are ever carried by railroad or other long-distance method of transport. They are hauled to the adjacent towns and cities and are used locally. They are hauled on the highways. The first movement from the farm is over the country roads and the cost of transportation from the farm to the station or to the local point of consumption may easily be as much as the freight cost to a remote market because of the small load that can be carried. The character of the rural highways plays a large part in the marketing of produce, and also in returning supplies to the farm.

The average haul of farm products to market by wagon was found by a special investigation by the United States Department of Agriculture to be seven and three-tenths miles. The average cost a ton to get to market for eleven important crops was one dollar and ninety cents. The cost a ton for a mile of transportation was twenty-six cents. The cost of railroad freight a ton mile was at the same time from one to two cents. Both figures would be much higher at this date. These facts emphasize the importance of rural highways in the problem of getting produce to market. The length of haul, the cost a ton and the cost to the ton-mile in New York as determined by the United States Department of Agriculture in 1907 were as follows:

To counteract the tendencies, many remedies have been proposed. Institutional features have been introduced. Particularly, agricultural and home economics lectures have been installed. Classes, clubs and social bodies have been organized, such as the brotherhoods at Wolcott, Cuba, North Collins, Bellona, Yorktown and many other places. Athletic equipment and industrial training have been installed at some places, agricultural fairs have been held as at Paris Hill and Cazenovia, the latter coupled with contests and athletic games. The rural Young Men's Christian Association at one time took a hand, as in Dutchess County, and installed a county secretary. Surveys have been made in particular regions to study the conditions and the problem.

CHAPTER XI

OUTLOOK

THE outlook for the agriculture and country life of New York is determined by its position, contributory population, and topography.

In position and population it occupies a commercial center. We cannot foresee any great shift in the commercial relationships, as harbors, waterways, natural lines of communication, do not change. The State will be called on for additional supplies to feed and clothe immense local populations. The populations will increase. The open country will be constantly invaded by these vast urban interests, and a special kind of rural life will develop, in close touch with the movements and affairs of large centers of population.

The natural diversity of the State, insured by its topography, requires that a wide range of products will continue to be grown. We cannot predict an Empire State predominantly devoted to fruits, vegetables, dairy products, or other particular lines. Special types of farming will undoubtedly increase in importance, to meet the needs of near-by populations, such as the growing of flowers, ornamentals, classes of vegetables and other products for which sufficient consumption can be secured only in a thickly settled region.

It is not to be expected that super-intensive farming will develop over very large areas inasmuch as great parts of the State are rough and elevated and must remain in relatively extensive agriculture. Although gardening will make great progress, for any future that we can foresee the State will not become a garden. It is not until other parts of the country fill up that the hills will be brought into any such state of cultivation as in parts of Europe; and even then it is doubtful whether such a development will take place, for human labor is to play a very different part in the agriculture of the future. Social conditions in this country, where there is no traditional attitude of master and man, will not allow of slavery to a piece of land. Probably many of the methods of raising the world's supplies will change radically within a century, and naturally these changes must come in the newer countries of the earth and where there is the greatest freedom of individual action.

New York has now had a long period of agricultural development. The commercial position of the State has taken on the characteristics of relative permanency. The general present rural situation, therefore, so far as farm products is concerned, is probably itself a fairly good forecast of the future. The State will never be predominantly agricultural. The industries and interests will have a coördinating and, let us hope, a coöperating evolution.

The character of the urban development will profoundly influence the course of the rural life of New York. Indications are that there must be a limit to population of great cities, that the smaller places will show a relatively increasing development. It is beginning to be realized that it is uneconomical to transport heavy supplies into a few great cities and to herd workers in them and then send the completed products back again whence the materials came. Thousands of carloads of print paper go into New York City to be sent out again to the west and south and north. Probably there will be a tendency for manufacturing plants to move to positions of more equal distributing ability and of less pressure. There will be a gradual urbanizing of the country, and a counter effect of ruralizing the cities. At present there is likely to be contest between urban and rural communities, but essentially these interests are common and the force of events must bring them closer together. Otherwise there is no prosperity for either.

Undoubtedly there will be a larger development of what may be called state farming, to utilize areas that cannot respond to private enterprise and from which the returns are matters of more than one generation. The present theory of utilizing swamp areas is to drain them in order to make them hard land for the tilling of ordinary farm crops. It is not unlikely that some of them can be better utilized in the future for the growing of swamp crops, as in some other countries, and for making water areas for the rearing of food fish. In either case, the utilization of vast areas now waste will require the powers of the State.

The present situation as to the scarcity and high

price of paper calls attention to the great areas that could be utilized for the growing of wood-pulp timber. Probably the eastern states could raise their usual supply of wood for paper, and improve their territories at the same time. There is needed a policy of state and county coöperation, that will bring thousands of acres into a forest agriculture for the production of paper stock and other wood supplies aside from the growing of building timber.

The rearing of wild life is a part of agriculture in the large sense. The protection of game is now involved partly in local laws backed by the pressure of sportsmen. There is a general antagonism between the farmer and the sportsman or hunter. A consistent state policy for the protection and rearing of all wild life, in areas adapted to it, in which farmers and all other citizens coöperate, could produce much in the way of supplies and at the same time conserve the sportsman and naturalist sentiment.

Fear of competition with the agricultural West has passed, now that consumption has relatively overtaken production and as we come to realize that all the national domain is to be needed for constructive productive uses. The agricultural position of New York is impregnable, and the farmers have confidence in its future, although discouraged at times, as are other farmers, by insufficient labor and by handicaps beyond their control. In fact, there has been a distinct return reaction from the West, in recent years, due to the cheaper land in the East.

We are to look for a moving down from the farther

and poorer hills, the organizing of farm units on a more economic and less patriarchal basis, and probably a still further increase, in such regions, in the size of holdings. With the spread of education, the use of more machinery, the necessity for closer cost-accounting, the facilities for disposal, the farmer's ability and power to produce supplies are increased. He can handle a larger investment than formerly. Consequently the older type of farmer will tend gradually to disappear.

THE APPROACH TO CONSTRUCTIVE POLICIES

Changes in rural organization and in public policies affecting agriculture are imperative. In approaching the subject, three primary human relationships are to be recognized. These are the principle of specialization in industry; the principle of coöperation or working together; the principle of democracy, or the expression of what is best in the individual.

The policies of constructive organization resolve themselves into about eight main divisions: 1. the land, and policies for its control; 2. weather and climate in relation to agriculture; 3. crop improvement; 4. live-stock improvements; 5. marketing, manufacture and distribution; 6. rural educational and social facilities; 7. investigation and research; 8. administrative and regulatory functions. It is not the purpose to enter into an extended discussion of any of these subjects here. The reader will have caught a point of view on most of them in the preceding pages.

Enough may be said, however, to make the meanings plain.

1. In considering the land and its control, an adequate survey of the surface of the State is the first step. The State is already well advanced in this respect through the work that has been done in coöperation with federal agencies in the topographic, the hydrographic and the soil surveys. These are underlying inventories of resources. Back of the soil and reaching into other lines is the geological survey. These should be pushed into the realms of chemical as well as physical study.

The development of regional field plot experiments on the more important types of soil is of paramount importance. In this respect New York has been very neglectful and is much behind other states.

Land utilization is here applied to the use of those areas that are not suitable for management in the ordinary type of farm where cultivated crops are the main product. This includes two types; thin rough land, perhaps remote from the railroad where some kinds of plants may still be grown without much individual attention. More than half of the area of the State is in this group. It is commonly considered suitable for forest, not because it will grow the best forests, but by a process of elimination of other crops. Wet land and areas of water remain to be developed as aquatic farms where many kinds of food can be produced in amounts, it is said, equal to or larger than are produced on similar areas of land farms.

In the land policy, it is important to recognize that a region of such diversified soils as New York will always have land on the margin between that suitable without question for tilled-crop farms and that adapted only for natural growth such as forest. This in-between land represents the economic shore line. In New York there are considerable areas once farmed that are now nearly idle and should remain so. The profit from their cultivation has fallen away. Some of those lands will again be included in farms. Other areas will continue, at least for a long time, to be best suited for other than ordinary farming purposes.

Land tenure and farm units are subjects of much discussion. The question is not primarily who owns the land or in what sized units it is operated, as that the land be well utilized and that the best possible farm people are produced.

The whole subject of taxation as it applies to rural property needs consideration in special relation to land.

2. Weather and climate in relation to agriculture need much further investigation. Weather observations and forecasting are aimed more at the facility and safety of shipping than at better agricultural production. In New York State the weather, meaning the longer-period average of weather conditions, is exceedingly diverse. As on overhead condition,

it often overrides the influence of the type of soil on crop production. This is well illustrated by the distribution of fruit and of certain special crops such as beans and hops. Both of these latter require a fairly rich calcareous soil, but beans need a dry harvest season. Hops do not require so dry a harvest time. Their distribution within soil regions follows lines of rainfall. Systematic investigation of local climate in relation to agricultural production should be instituted.

Crop ecology, or the relation of important varieties of plants to all the external conditions under which they will thrive, is also in need of more attention.

3. The crop, quite as much as the soil, may be the object of improvement for larger production. Regional adaptation of varieties has not been worked out thoroughly. The present distribution is the result of unconscious adjustment by farmers to the results of their own experience. This of course has value, but something can be done by systematic study by the field survey method to set down the limits of this adaptation. A broad view of environmental factors, such as soil, climate and topography, as well as of market conditions should be taken.

More uniformity in the varieties of crops grown in a region will not only simplify the problems of production but will also aid in marketing, since larger lots and better grading can be secured. These have a definite advertising value.

Crop insurance is practicable only within limits. It can be applied with most success where those natural forces are involved against which protection is not available, such as hail and flood. It is scarcely feasible, however, to have insurance against such sources of loss as oats smut, San José scale or coddlin-moth of the apple, since there are well known remedies for these and insurance would in large measure put a premium on sloth.

4. The keeping of live-stock should generally render it easier to maintain the productiveness of the soil than crop husbandry alone. The animal manufactures plant products, some of which may be unsuited to human food, into usually more palatable animal products. If care is taken to make up the deficiency in the soil resulting from the lack of animal manure by the use of commercial fertilizer, lime, and the not too close removal of crops so that the humus content of the soil is maintained, it is quite as possible to maintain productiveness of the soil by crop as by animal husbandry. The higher the prevailing price of human nutriments, the less desirable is live-stock husbandry. The natural swing in New York should be toward a larger proportion of crop products and a less proportion of animal products. Vegetables, fruits and flowers should increase in importance relative to coarse forage and grains. The large area of low-grade range and forage land will continue to hold live-stock husbandry in many regions.

The motor vehicle is forcing a certain readjustment in crop distribution. New York has led in the production of market hay, chiefly timothy, which has been sold in the cities for the consumption of horses. The development of the gasoline motor is rapidly reducing the demand for hay. The question is whether the farmer can feed this hay to other types of live-stock or whether he will be obliged to change his system of farming. A reduction in the acreage of hay seems desirable, especially on land that can be used equally well for other feed crops.

The question may be asked as to what extent it is legitimate for the farmer to be paid for the depletion of his soil by cropping. The limit of such payment should be the extent to which the soil is permanently reduced in productive capacity. It is measured by the amount of commercial manures and fertilizers that are necessary on the average to maintain the yield on the poorest land farmed.

Regional adaptation and types of live-stock are quite as important as uniform crop varieties. In New York most of the animals kept on farms, aside from the horse, are on a by-product or waste land basis. That land, therefore, largely determines the cost of producing a particular type of animal product. Cultivated land of poor quality has the largest difficulty in competing with such production.

The question may be raised whether the manufacture of evaporated, powdered and condensed milk will affect the market milk business in New York State. The swing from cheese and butter to market milk, which has been large in recent years, has been due to the embargo in the form of freight charges placed on milk from distant states, together with its perishable character. But if evaporated milk can be produced that can be handled as a staple commodity, and if milk powder can be revived with all the physical and chemical properties of fresh milk, it may take away the special advantage of New York in the dairy business. Milk is produced cheapest where feed and forage are cheapest and on the whole this is recognized to be in the Middle West.

5. In marketing, manufacture and distribution, the following subjects are of outstanding importance: Conformity on the part of farmers to a normal unit of shipment, which is usually the carload; establishing of standard grades of farm products; storage for seasonal products; effort to put the products of the farm in such preserved and manufactured form as to make them stable articles of commerce the year round; publicity on the visible supply and current prices for all food products by grades; encouragement of community assembling and packing facilities as the starting-point of the handling of farm produce. The ultimate object in all plans is to insure to the producer a fair return on his effort and to the consumer a reasonable price. The taking of toll by a large number of intermediaries reveals the crude nature of the present practices in distribution and selling, and works injustice at both ends of the line. Public oversight of many of these practices is a necessity if justice is to be accomplished.

6. The rural educational and social facilities need great extension, particularly in these lines: vocational training in the grades as well as better teaching generally in the rural schools and in larger units;

the special schools of agriculture should become more organically a part of the state educational system and gradually correlate with the high-schools; the extension type of vocational instruction which has attained large development in the farm bureau system should be extended, and the home demonstration agent be constituted a regular part of it; every community should study its own social facilities through the Grange or other recognized agency. The social functions in the open country should be considered to be definite and worthy of separate development and not dwarfed or overshadowed by business or religion. This function should preferably express itself through one body. The smaller the community, the greater is the necessity that it confine these activities to one organization.

7. Investigation and research should always be the subject of encouragement by a government. A growing population is continually pressing against the bounds of its established knowledge. Agriculture has been one of the first divisions of business to set up investigation and research as a distinct public function to which men exclusively devote their time. The New York Agricultural Experiment Station was probably the first of its kind in this country to be organized on a comprehensive basis.

The tendency of routine government organization to crystallize into easy rather than altogether efficient lines of administration brings danger of stifling that essential freedom of thought and action which is the essence of effective investigation. The public and its legislative representatives need to understand that such persons and such institutions cannot be administered successfully on the same basis as asylums and penal institutions. A reasonable generosity in the management of funds is essential, and those responsible for investigation should be given the utmost freedom in the use of the funds. The accounting should come at the end, instead of being laid out in literal detail in advance.

Better organization is needed of the natural-resource studies. There is need of a systematic effort, through conference and coöperation, to bring all these agencies into touch and to assemble, analyze and put into the most usable form all the information that is being collected. Such a coördination might be called a Natural Resources Survey, whose most definite antecedent is the conservation movement. It might be given permanent status in a State Academy of Science.

8. It may be worth while to point out the desirability of keeping the administrative and regulatory function separate, as far as possible, from the educational and investigational function. The executive departments that enter into the Governor's cabinet should not be permitted to encroach on the educational and investigational functions. The Council of Farms and Markets has very closely attained that status. As a principle of government especially applicable to the Council of Farms and Markets, it should be recognized that more responsive and better service will result from responsible individual heads

of departments of cabinet rank and carrying authority commensurate with the responsibility, than from the Commission type of head. The most important regulatory function to be stressed is that of grades and standards of farm produce.

Closely associated with grades and standards is the collection of "up-to-the-minute" information on the supplies and prices of important commodities and their publication together, in such form that the public will understand the essential facts and learn the coördination between them. Supplies need to be reported from all sources including all public and private storage and warehouses. Ignorance of supply is the largest factor in protecting speculation.

Finally, a word may be added in appreciation of the large agricultural resources and achievements of the State. Very much has been accomplished, but there are large possibilities for future growth through better adaptation of crops and live-stock to local conditions, a better understanding by the farmer of the possibilities that lie within his own achievement, and through the improvement of rural institutions chiefly in details of operation that rest largely on the human element. The past generation has been the development of physical resources; the great problem now before us is the improvement of the human population.

New York agricultural people may look forward to a growing condition of rural prosperity. The present swing of events is in their direction in contrast to the swing against them in the period of thirty to forty years ago. We have seen that, although the State is old as measured by American history, it is new as measured by the history of civilization; and we have learned that its resources in an agricultural way are yet largely to be developed.

As we began this book with the motto on the Great Seal of the State, so may we close with it if we interpret "Excelsior" to mean, "More is to be achieved."

INDEX

A

Abandoned farm region, 96
Acid soils, 94
Adams Act for research, 337
Adirondack Mountains, 19
  soils, 110
Administrative functions, 367
  organization, 293
Agrarian movement, 61
Agriculture, Board of, 81
Agricultural decadence, 76
  department, origin, 82
  development, 141
  development, best, 97
  divisions, 300
Agriculture in High Schools, 322
Agricultural institutions, 81
Agriculture, New York Society for Promotion, 81
Agricultural Society of New York, prizes, 80
  origin of, 349
Agriculture, State College of, 323
Agriculture in Rensselaer Polytechnic Institute, 83
Albany lumber center, 119
Alfalfa, 144
  best soils for, 99
Animal Industry, bureau, 301
  early, 200
Animals as food producers, 202
Animal production, intensity of, 201, 203
Apple juice, 289
Apples among Indians, 60, 175
  evaporated, 179
  first shipped abroad, 176
  grades, 178
  leading varieties, 178
  new varieties originated, 176
  production of, 172
Apples, storage and manufacture, 179
Area in farms, 38
Area, land and water, 3
Ashes, sale of, 74
Automobile truck lines, 69
Ayrshire cattle, 218

B

Babcock milk test, 279
Bailey, Liberty Hyde, 86
Barbour soils, 115
Barge Canal, 34, 67, 251
Barley, 157
Barns, dairy and storage, 216
Barrel manufacture, 286
Barry, Patrick and nursery development, 193
Basic rocks, 18
Beach bars of gravel, 26
Beans, dry, 165
  and sheep raising, 167
  soils, 99
  in wheat culture, 156
  on farms, 243
Bees, foul-brood in, 303
Berkshire Hills, 12
Big Tree, Geneseo, 59
Bird life, 128
Birds, protection of, 129
Black ash salts, 119
Blackberries, 184
Blue-grass, 150
  soil relations, 101
Bostwick, Reverend William, and grape growing, 187
Boweries, farms, 56
Breeders' association, 352
Brewer, William H., 84
Brown, Amos, 84
Brussels sprouts, 170
Buckwheat, 159
Buffalo, reason for location, 2
Bulletins, agricultural, 338
Bush-fruits, 184
Butter factory, first, 278
Butter, oleomargarine, 302

Butter production, 220, 278

C

Cabbage, 167
  soils for, 99
Calcarious soils, 101
  soil in pasture, 151
  for beans and peas, 166
Calciferous-Trenton formation, 20
Calquahan, Patrick, 72
Canada blue-grass, 151
Canal, Barge, 34
  boats, capacity and rates, 67
  Champlain, 34
  Chenango, 35
  Deleware and Hudson, 34
  early, 65
  systems, 34, 251
Caneadea soils, 115
Canning industry, 290
Carboniferous rocks, 16
Catskill formation, soils from, 94
  group of rocks, 23
  height of mountains, 5
Cattle, beef, 221
  early types, 216
  pure blood, numbers, 217
Cauliflower, 170
Celery on muck soil, 169
Cement production, 133
Central Lakes, rainfall, 54
Certified milk, 277
Champagne, production of, 191
Champlain Canal, 34, 251
  discover, 56
  valley, features of, 5, 7
  soils, 104
  apple region, 175
Chapin, Mr., 176
Charcoal production, 74
Chautauqua grape belt, 188
Chenango Canal, 35
  soils, 96
Cheese, production of, 220, 279
Cheese factory, first, 280
  system of dairying, 215
Cherries, 181
  sweet, 182
Chickens, production of, 238
Church in community life, 352
Cider making, 288
Clermont, Fulton's boat, 66
Clays, for ceramics, 134
Climate and crop distribution, 54
  for grape growing, 189
  of New York, 41
  need for local study, 361
Clinton, Dewitt, and agriculture, 81
  Robert, 83
Clinton formation, 21
Clover production, 144
  seeded with timothy, 156
Clyde soils, 103
Cloudiness, 53
Clydesdale horses, 226
Coach horses, 226
Coal age, rocks of, 16
Coastal plain soils, 111
Coloma soils, 105
College, New York State Agricultural, 324
Columbia University, 332
Commission Merchants' bond, 308
Condensed milk, 264, 281
Confederacy of Indians, 58
Conservation commission, 309
Continental glacier, 24
Coöperative Associations bureau, 306
Coöperative cheese factory, 280
Cooledge, James D., and hop production, 195
Collier, Peter, 86
Cooperage products, 286
Corduroy road, 66
Corn destroyed by General Sullivan, 59
  for grain, 152
  silage, 149
  sweet, 168
Cornell, Ezra, 324
Cornell experiment station, 336
  soil of, 103
Cornell University, origin of, 84, 324
Counties, origin of, 71
Country population, 40
Cover crop of buckwheat, 160
Cover crop of rye, 158
Cows, dairy, number, 207
Cows, milch, distribution, 212
Cranberries, 185
Creamery, first, 278

Credit facilities, 297
  facilities needed, 270
Cretaceous formation, 27
Crops, change in area, 75
  cultivated area, 77
  distribution and climate, 54
  ecology, need for study, 362
  of Indians, 59
  industries, 138
  rank in production, 140
  trend of yields, 78
Cucumbers, 170
Culver soils, 96
Cumberland plateau, 12
Currants, 184
Curtis, Col. F. D. and Duroc swine, 238

D

Dairy barns, 216
  bureau, 302
  cows, 212
  cows, number, 207
  herds, size of, 218
  manufacture, 275
  sanitary measures, 277
Dairying, influence on social development, 212
  and soils, 102
  summer, 214
  winter, 214
Dairymen's Association, 352
  League, 88, 262
Decadence, Agricultural, 76
DeKolb soils, 94
Delafield, John, 79, 83
Delaware and Hudson Canal, 34
Deltas, glacial, 26
Devonian series of rocks, 23
Dewitt, Simeon, 83
Dimension of State, 3
Discovery of New York, 56
Dogs, menace to sheep, 230
Dover soils, 105
Downing, Andrew Jackson, 193
Downing Nurseries, 190, 192
Drainage, legal provisions for, 311
  surface, on Vergennes soils, 108
  systems of state, 31
  tile, rise of, 79
Drumilins, 27
  Wayne County, 9
Drying fruit, 273
Ducks, production of, 242
Dunkirk soils, 102
Dunn, C. and Cotswold sheep, 233
Durham boats, 65
Dutch belted cattle, 217
Dutch Settlements, 60, 63
  West Indies Company, 56
Dutchess of Geneva, Eighth, Cow, 216
  soils, 109

E

Economic pressure, 13
Education, Department of, 314
  needs, 365
  organizations, 314
Eggs, production of, 239
  record production, 242
  type preferred, 241
Electric railroads, 250
Empire state, 58
English settlement, 63
Ensilage, pea, 167
Erie Canal, 34, 65
  enlargements, 66
  Lake, plain, 10
Evaporated apples, 179
  fruit, 273
Extension bulletins, 339
  movement, 86
Experiment stations, 335
  origin, 85
  soils, 103
Extension teaching, 339

F

Fairs, Agricultural, early, 81
  County and State, 351
  rise of in State, 82
Falls of Niagara, 22
  water, 33
Farms, abandoned, 96
Farm area, 38
  bureau movement, origin and function, 342
  loan system, Federal, 298
  and Markets, Council, 299
  organization, relation to motor vehicle, 363

Farm area (*continued*)
  organization, relation to powdered milk, 364
  population, 40
  for sale, list of, 304
  settlement bureau, 304
  woodlot, 120
  tax exemption, 127
Farmers' institutes, 340
  movement, 86
Fay, Elijah and grape growing, 186
Fay, Lincoln, grape nursery, 193
Fayette Agricultural School, 83
Female suffrage among Indians, 58
Fertilizer inspection, 302
  manufacture, 285
Fertilizing muck soil, 169
Feudal tenure, 70
Finger Lakes, 10, 31
  and local storms, 53
  region, 25
  soils, 99
  and rye, 158
Fire patrol area, 120
  regulations in forests, 125
  protection of forests, 125
First bottom soils, 114
Fish, 128
  culture, 130
  and game division, 309
  hatcheries, 130
Fisheries, shell, 131
Floods, 54
Flour, early production, 75
Flower City, Rochester, 75
Flowers, production, 194
Food inspection, 308
Food and markets division, 258, 304
  products bureau, 308
  production by different animals, 202
  values per acre of crop, 163
  waste of, 269
Forage crops, 142
Forest area, 77
  areas, 120
  fire protection, 125
  management, 123
  nurseries, 127
  policy of state, 124, 357
Forest area (*continued*)
  regulations, private, 127
  types, 122
Forestry, instruction in, 127
  state college of, 332
  and the state constitution, 124
Fort Orange, 56
Foster, William, introduced Merino sheep, 233
Fredonia, site of first grange, 87
French settlers, 61, 63
Frost, danger of, 49
  free period, 47
  and fruit production, 175
Fruit belt, 174
  dried, 290
  evaporation, 273
  packing houses, 178
  in Hudson Valley, 110
  soils for, 99, 103
Fruits, 172
  changes in area, 78

G

Game, 128
  farms, 129
Gas, natural, 133
Geese, production of, 238
Genesee, reservoir site, 33
  river, 32
  road, 66
  soils, 115
    for celery, 170
  tract, 72
  valley and wheat, 156

Geneva agricultural experiment station, 85, 103, 336
Geological structure, 14
Geology of New York a landmark, 17
German settlers, 62
Ginseng, 197
Glacial influence on agriculture and travel, 29
  lake soils, 102
  lakes, 25
  soil formation, 90
  till soils, 98
Glacier, continental, 24
Glass house farms, 195

Gloucester soils, 109
Glove manufacture, 284
Goats, as milk animals, 235
Gooseberries, 184
Gorham, Nathaniel, 72
Goshen butter, 278
Grain crops, 152
  soils, 101
Grange, functions and influence, 348
  origin in New York, 87
  League-Federation exchange, 268
Granite quarries, 132
  rocks, 18
  in soil, 98
Grants of land, 60
Grape Belt, 187
  juice, 187, 289
Grapes, Chautauqua district, 188
Grapes, extent of production, 191, 192
  first car shipment, 187
  Hudson district, 189
  Keuka districts, 187
  leading varieties, 186, 191
  Niagara district, 190
  rise of production, 186
Grass and pasture area, 77
Great Lakes frontage, 30
Great Seal, motto on, 1
Green forage, 148
Green-manure of buckwheat, 160
  of rye, 158
Green vegetables, 171
Growing season for fruit, 175
  length, 47
Guernsey cattle, 218
Gypsum beds, 22
  production, 134

H

Haight, D. H. and Dutch belted cattle, 217
Hail, occurrence, 53
Hamilton Shales, 23, 99
Hanging valleys, 24
Harbor Moraine, 111
Hardpan soils, 96, 98
Hatch Act for experiment stations, 335
Haul, lengths of, 253
Hay and forage crops, 142
  pasture area, 203
Heaton, Mr. and shorthorn cattle, 216
Heft, Peter Earstin and early live stock, 200
Herkimer, General, 71
Helderberg limestone, 22
Hempstead Plains, 112
Highways, state system, 254
Highland areas, 11
Hinson, W. O., and tile drainage, 80
Hired men on farms, 40
Hodenosounce Indians, 57
Hogs, production of, 236
"Hogsbacks" hills, 10
Holland land company, 70
  origin, 72
Holstein dominant dairy breed, 217
  Friesian cattle, 200
Home Economics, 330
Honey, production of, 243
Honeoye soils, 100
Hop production, 195
  soils, 99
Hopper, H. A., sheep for New York, 234
Horses, 222
  production and consumption of in state, 224
  shows, 226
Horticultural Society of New York, 88
Hornby, William, 72
Houghton Farm, First experiment Station, 85
Howes, Sidney and Berkshire swine, 238
Hoxie, Solomon, established advanced register, 218
Hub City, Syracuse, 246
Hudson, Hendrick, 56
Hudson River formation, 20
  formation and soils, 105
  first steamboat on, 66
  tide in, 57
Hudson Valley, apple region, 174
  dairying in, 213
  features of, 5
  gorge under sea, 6, 30
  soils of, 108
Huguenots settlements, 61

Humidity, 53
Hungarian, grass, 148
  settlements, 63
Hunting license, 310

I

Indian agriculture, 57
Indian Castle, 59
Indian plantings of apples, 175
Industry, patrons of, 349
  and physiography, 37
Infectious disease, regulation, 301
Institute, Farmers', 86
Institution, farm bureau, 304
Institutions of Agriculture, 81
Investigations, need for support, 366
Irish settlements, 63
Iron ore, 131
Iroquois Indians, 57
Italian settlements, 63

J

Jersey cattle, 218
Johnston, John, 79
Johnson, Sir William, 63
Jordan, W. H., 86

K

Kelley, O. H., 87
Kentucky blue-grass, 150
Killing frosts, 48
Kraut manufacture, 168

L

Laborers, farm, 40
Lackawanna soils, 96
Lake, glacial soils, 102
Lakes apple region, 174
  glacial, 25
  plains' soils, 97
  region grape district, 188
  soils, 107
Land, area, 3
  changes in productive capacity, 78
  grants, 60
  Grant Act, relation to agriculture, 324
  plaster, 134
Land, area (*continued*)
  prices, early, 72
  tenure, 361
  utilization, 360
  values, 296
Latitude of New York, 2
League, Dairymen's, 88
Leases, limits on, 70
Lettuce, 169
Lexington of North, 222
Liberty, statue of, 2
Library, State, 315
Licensing commission merchants, 257
Lima beans, 170
Lime deficiency in plateau soils, 94
  for Long Island soils, 114
Limestone abundant in soil, 100
  in Ontario soils, 99
  produced in state, 132
  in soil, 97, 105
  source of in soils, 91
Linnean Botanical Garden, 192
Little Falls dyke, 8
Live stock, as food producers, 202
  of Indians, 60
  place in state, 202, 363
  relation to soils, 99
  sales, 264
  shifts in numbers, 205
  and soil fertility, 209
Livingston, Chancellor, 61
Livingston, Robert, first barrel apples abroad, 176
  and early Merino sheep, 233
Local storms, 52
  winds and breezes, 45
Lockport shales, 22
Long House of Indians, 57
Long Island, features of, 11
  soils, 111
  structure, 27
Longitude of New York, 2
Lordstown soils, 96
Lowland area, 11
Lumber center, Albany, 119
  early production, 75
  industry, 285

M

McCombs, Alexander, 73

Manufacturer's rural, 271
Manhattan Island, 56
Manors, English, 69
Manure, production of, 209
  for soil improvement, 209
Maple sugar production, 199
  and syrup, 288
Marble, occurrence of, 132
Marcy, height of mountain, 5, 12
Marine swamps, 115
Marketing, agencies, 257
  some principles, 260
Markets and marketing, 245
Markets, municipal, 259
  standards and units, 365
  and storage bureau, 305
Marl deposits, 28
Martin, Jonas, shipped grapes, 187
Massachusetts preëmption right, 63
Meat packing, 272
  industry, 282
Medina formation, 20
  sandstone, 105
  shales, 99
Men on farms, 40
Merrimack soils, 108
Milch goats, 235
Military tract of Massachusetts, 71
Milk, certified, 277
  powder, 282
  powder, relation to farm system, 364
  production per cow, 219
  quantity produced, 276
  trains, 263
  train, first, 213
  use of, 220
Millet, 148
Milling business, 291
Mills, grist and saw, 75
Minerals produced, 132
Mines, 131
Minuet, Peter, 56
Mohawk soils, 101
  valley, features of, 7
  soils, 97
Molding sand, 134
Morain, Harbor, 111
Morrill Act of 1862, 84
Morrill, Senator Justin, 84
Morris, Robert, 72
Motor vehicle, effects on type of farming, 363
Motto on great seal, 1
Muck deposits, 28
  soils, 116
  for cabbage, 167
  crops, 169
  onions on, 169
  for peppermint, 198
Mules, 227
Museum, State, 315, 323
Mushrooms, 170

N

Narrows, 30
Natural History, of New York, report on, 17
  survey of New York, 82
Natural Resource Survey, need for, 367
Nature study and bird life, 129
  in schools, 320
Needham, J. G., 130
New Amsterdam, 56
Newberry, glacial lake, 26
Niagara formation, 22
  River, cascade, 10
Nixon bill for agricultural extension, 346
Nixon, S. F., and agricultural extension, 346
Norite rocks, 18
Nurse crop of buckwheat, 160
Nurse crop, oats, 155
Nurse crop of wheat, 156
Nursery development, 192
Nurseries, forest, 127
  inspection, 303
Nuts, production of, 185

O

Oak openings, 74
Oats, 154
Oleomargarine, regulation of sale, 302
Oneida lake, features, 31
Oneonta, formation, 23
Onguie-Honwe, 57
Onions, production of, 169
Onondoga Castle of Indians, 59
Ontario Lake plain, 9
  temperature of water, 48
Ontario soils, 99

"Orange County red" onions, 169
Orchard cover crop, buckwheat, 160
Orchard grass, 148
Orchard inspection, 303
Organizations, agricultural, 347
Outlook on New York agriculture, 355
Ovid Academy, 83
Oxen, place in state, 221
Oyster beds, 131

P

Packing houses, central, 265
 fruit, 178
Palisade region, 6
 elevation of, 12
 formation of, 18
Palatines, 62
Panama rocks, 16
Papakating soils, 115
 for celery, 170
Parmenter, Mr., and grape growing, 186
Pasture, 150
 and hog area, 202
 value of, 204
Patrons of Husbandry, 87
Patroons, Dutch, 69
 settlements of, 61
Peaches, area in, 179
 districts growing, 179
 leading varieties, 180
Pearl ash, 74
Pears, leading varieties, 181
 production of, 180
Peas, dry and green, 165
Peat deposits, 28
Peat soil crops, 169
Peoples' College, 84
Peppermint for oil, 198
Physiography of State, 4
Percheron horses, 225
Petroleum, 133
Phelps, Oliver, 72
Physiography and industry, 37
Plains, Hempstead, 112
Plank road, 66
Plant industries, 138
 bureau of, 303
Plateau region, climate, 49
 soils of, 94
 for buckwheat, 160
Plateau, southern, 12
Plums, 181
Polish settlements, 63
Political divisions, 294
Population, center of in New York, 246
 distribution of, 37
 Indian, 59
 rural, 40
Portages, early, 65
Portage, reservoir site, 33
Potash, early production, 74, 119
Potatoes, production, 142, 161
 soils for, 106
Potsdam sandstone, 19, 105
Poultney, Sir William, 72
Poultry, types of breeds, 241
 production of, 238
Powdered milk, 282
Prairie tracts, 74
 land, 113
 Hempstead, 118
Precipitation, 50
Preëmption right of Massachusetts, 63, 72
Protein production per acre, 163
Protestant settlements, 62
Prince nurseries, 192
Prunes, 181
Public Service Commission, 252
Puritan settlements, 62

Q

Quack-grass, 168
Quaker settlements, 62
Quarries, 131
Quince production, 183

R

Rainfall, 50
 abnormal, 54
 and bean production, 166
Railroads, early routes, 68
 electric, 68
 mileage, 35, 249
Rank of state in crop production, 140
Raspberries, 183
Reading course bulletins, 339
Red soils, 96, 115

Redtop grass, 150
Regents, Board of, 315
 system of instruction, 319
Rensselaer Polytechnic Institute and agriculture, 83
Rent war, 70
Research, need for support, 366
 organizations, 314
Reservoir sites, 136
Richard's vineyards, 186
Ridge road, 26
Roberts, Isaac P., 85
 and tile drainage, 80
Rochester, Flower City, 75
Rock City, 16
Rock salt, 133
Roads, and topography, 36
 early routes, 65
 public, mileage, 254
Root crops, 148
Rose, Dr., and Southdown sheep, 233
Rotation in bean region, 166
Routes of travel, 32, 65
Runoff of rainfall, 135
Rural, development, 355
 property, assessed value, 295
Rye, 157
Rye with vetch, 159

S

Salina formation, 22, 99
 salt production, 133
Salt, beds, 22
 City, Syracuse, 22
 production, 132
 springs, 22
 rock, 133
Sand, molding, 134
Sandstone, occurrence, 132
Saw-mills, 292
St. Lawrence Valley soils, 104
Schools of agriculture, 330
Schools, early agricultural, 83
Secondary education, 321
Science, State Academy of, 367
Scotch settlements, 63
Sea breezes, 45
Seal, motto on great, 1
Seneca Castle of Indians, 59
Seneca Turnpike, 66
Series of soils in state, 94
Settlement, lines of, 57
 progress in, 60
Sheep, changes in number, 228
 history of introduction of breeds, 233
 leading breeds, 234
 raising, effect of competition, 230
 relation to dairying, 232
Shipments in early days, 65
Shape of State, 3
Shell fisheries, 131
Short courses in agriculture, 328
Shorthorn cattle, early prominence, 216
Silage, corn, 149
Silo, place in dairying, 215
Six Nations of Indians, 57
Slate occurrence, 132
Slaughtering, on farm, 282
Slocum, David, College of Agriculture, 332
Small-fruits, 183
Smith-Lever Act for agriculture, 345
Snook, William, introduced teasel, 196
Soap manufacture, 284
Society for promotion of agriculture, 81
Southern plateau, 12
Soils, best, 97
Soil, changes in productive capacity, 78
Soil, fertility and animal manure, 209
Soils, first bottom, 114
Soils, formation of, 90
 depth of, 93
 for potatoes, 164
 of New York, 90
 red, 96
 special, 114
 survey, area covered, 93
Soiling, 215
Soiling system, 159
Special soils, 114
Spencer, John, 87
 Uncle John, 320
Spray materials, inspection, 303
Springs occurrence, 33
Standards, legal, of foods, 257
 of marketing, 365

Starch production per acre, 163
Stanwix, Fort, treaty, 71
State College of Agriculture, 85
Statistics bureau, 304
Statue of Liberty, 2
Steamboat, Fulton's *Clermont*, 66
Stockbridge Hill, orchard of Indians, 60
Storage for apples, 179
Storage function in agriculture, 269
Storms, local, 52
Strawberries, 184
Sleet storms, 53
Sturtevant, E. L., 86
Suffrage, female among Indians, 58
Sugar, beets, 171
  bush, 198
  maple, 198
Sullivan's expedition, 63
  raid, 59
Summer boarder influences, 109
  dairying, 214
Sunshine and cloudiness, 53
Surface features, 4
Surveys, of natural resources, 360
  soil, area covered, 93
Swamp soils, 115
Sweet corn, 168
Swine, leading breeds, 238
  production of, 236
  relation to poultry raising, 237
Syracuse, Salt City, 22
  University, 332

T

Tadpole hills, 9, 27
Tanneries, 284
  early, 75
Tax exemption on timber land, 126
Taxation system, 295
Teasel, as a crop, 196
  soils, 99
Telephone system, 256
Temperature ranges, 47
Terminal morains, 25
Terraces, glacial, 91
  soils, 110
Territorial development, 69
Thousand Islands, 8
Thayer, Col. Jared, first cheese factory, 280
Thunder storms, occurrence, 52
Through valleys, 13
Tidal swamps, 115
Tide in Hudson River, 6, 57
Tile drainage, origin, 79
Till, glacial deposit, 29
  soils, 98
Timber, average stand, 121
  early growth, 74
  growth on plateau, 96
  on Worth soils, 107
Timothy production, 144
Tobacco production, 196
Tomatoes, 168
Topography, 4
  and roads, 36
Tory lands, 63
Tractors and horse breeding, 224
Transportation facilities, 248
Travel, routes of, 32, 65
Tree growth, 118
Trenton limestone, 100, 105
Triassic rocks, 18
Tribal Council of Indians, 58
Trotter's, American, 226
Trotting horse, notable animals, 222
Trucks, motor routes, 256
Tryon County, 71
Tuberculosis, payment for animals, 301
Turnpikes, 66
Turkeys, production of, 242
Types of soil, 94

U

Uncle John Spencer, 87
Under-drainage, rise of, 79
Underhill, Robert, and grape growing, 186
Unglaciated soils, 94
University of State of New York, 314
Universities, students in, 317
Utica shale, 101
Urban development, 356

V

Valentine, Lawson, 85
Valleys, form of slopes, 13
Van Alstyne, Edward, 86
Vance, L. J., on wine production, 192
Van Rensselaer, Killan, 61
Vegetables, 161
  area of, 171
Vegetables on muck soils, 117
Vergennes soils, 107
Vetch with rye, 159
Veterinary Medicine, Schools of, 333
Vinegar making, 288
Vocational education, 319
Volusia soils, 96

W

Walloons, settlement by, 56
Warren, G. F., on value of pasture, 204
  operation of freight as a tariff, 237
Warren, glacial lake, 25
Water area, 3
  falls, 33
  power, 136
  supply, 33, 135
  transportation, 251
Waterways, 29
Watson, Elkanah, 81
Weather, need for study, 361
  of State, 44
  service, 313
  origin, 45
Weeds, common in poor pasture, 151
Weevil, in wheat, 156
Weights and Measures Bureau, 307
Welch settlements, 63
Welland Canal, 30
West Indies Company, 56
Wheat, production of, 155
  competition in production, 65
  early cash crop, 75
  growing, competition with West, 157
  weevil, 156
White, Dr. Andrew D., 324
White clover, 150
Whitney, Joshua, 65
Williams, Jesse, coöperative cheese factory, 280
Willows, for basket making, 198
  ware, 287
Wind, direction of movement, 45
Wine-making, 186
Wine production, 191
Wing, Joseph E., on sheep in New York, 234
Winter dairying, 214
Women on farms, 40
Wood lot, farm area, 120
  tax exemption, 127
Wood pulp, 287
Woodland pasture, 203
Wool, production per animal, 228
Woolen manufacture, legislation promoting, 233
Wooster soils, 96
Worden, Schuyler, originated grape, 187
Worth soils, 105
Wright, Wheeler D., on bees, 243

Y

Young, Robert, and Cheviot sheep, 233